The Restless and Relentless Mind of Wes Jackson

THE RESTLESS AND RELENTLESS MIND OF WES JACKSON

Searching for Sustainability

ROBERT JENSEN

Foreword by David W. Orr

University Press of Kansas

© 2021 by the University Press of Kansas
All rights reserved

Published by the University Press of Kansas (Lawrence, Kansas 66045), which was organized by the Kansas Board of Regents and is operated and funded by Emporia State University, Fort Hays State University, Kansas State University, Pittsburg State University, the University of Kansas, and Wichita State University.

Library of Congress Cataloging-in-Publication Data
Names: Jensen, Robert, 1958– author.
Title: The restless and relentless mind of Wes Jackson : searching for sustainability / by Robert Jensen ; foreword by David Orr.
Description: Lawrence : University Press of Kansas, 2021.
Identifiers: LCCN 2020026495
ISBN 9780700630554 (cloth)
ISBN 9780700630561 (epub)
Subjects: LCSH: Jackson, Wes, 1936—Interviews. | Sustainable agriculture—United States. | Environmental responsibility—United States. | Botanists—United States—Biography. | Environmentalists—United States—Biography. | Geneticists—United States—Biography.
Classification: LCC S494.5.S86 J46 2021 | DDC 338.10973—dc23
LC record available at https://lccn.loc.gov/2020026495.

British Library Cataloguing-in-Publication Data is available.

Printed in the United States of America

10 9 8 7 6 5 4 3 2 1

The paper used in this publication is acid free and meets the minimum requirements of the American National Standard for Permanence of Paper for Printed Library Materials Z39.48-1992.

Contents

Foreword by David W. Orr   vii

Introduction: In the Barn   1

1. The Energy of Life   23

2. Agriculture   35

3. Taking Nature Seriously   43

4. Too Many, Too Much   55

5. Knowledge   75

6. Schooling   87

7. "We" Have to Accept Limits   97

8. A Creaturely Worldview   105

Conclusion: Spontaneous Elaboration   119

Appendix: Mission Statements of The Land Institute
from *The Land Report*   127

Notes   131

Published Works of Wes Jackson   139

A photo section follows page 65.

Foreword
David W. Orr

Wes Jackson is a big thinker, a generalist who asks questions that cross boundaries of disciplines, time, propriety, and almost everything else. In an age of academic specialization, he describes himself as a "herd quitter." Jackson, however, is no dilettante but is firmly grounded, figuratively and literally, in land, with its limits and possibilities. His life is a series of conversations about the rules of the "ecosphere" that govern soils, plants, animals, and human behavior. It's all of a piece, broken apart at our peril.

Jackson is also a gifted storyteller who likes people, good jokes, controversy, and conviviality. Combined with a restless energy, the result is a man, as Robert Jensen describes below, who takes on some of the aspects of an Old Testament prophet. Prophetic or not, Wes is the most adept and sane Bible quoter I know. A conversation with Jackson is likely to roam over religion, literature, philosophy, science, current affairs, human foibles, personal reminiscences, and quirky stuff he's noticed along the way—all of it wrapped up in memorable phrasing and tied together with the kind of humor you'd expect from a guy raised on a Kansas farm.

Over the years, much has been said about Wes, and I won't add to that here. Suffice it to say he is an "interesting" man, not just because of an outsized personality but also because he has made the foundational issues of humankind and nature interesting in ways few others have done. For Jackson, the story begins with the invention of agriculture ten thousand years ago this month. That inauspicious event opened a rift between humanity and nature that has grown into a chasm in the age of industrial, capital-intensive agriculture that presumes to improve a gift without gratitude or knowing much about what has been given. Although it certainly has technical aspects, the problem of agriculture is not fundamentally technical but rather epistemological, a fancy word to describe how we know what we think we know. The problem, as Mark Twain once said, is that a lot of what we know just ain't so. While acknowledging that we are much "more ignorant than knowledgeable," Jackson is also fond of saying that we should ask "questions that go beyond the available answers." Meanwhile, back on the farm, the question of how we will feed ourselves in a hotter, more capricious climate looms larger every year. Permanent answers are not likely to come from agribusiness and its allies.

By almost any measure other than corporate profit, American agriculture has been an epic disaster. I once asked my seminar students to map the relationships that explain the rates of obesity in the United States and the expanding dead zone in the Gulf of Mexico. As I watched, they began to fill the whiteboard with lines and arrows connecting farm policy, subsidies, fast foods, corporate farming, capital markets, giant agribusinesses such as Monsanto, petrochemical companies, equipment manufacturers, congressional committees, the US Department of Agriculture, advertisers, and so forth.[1] Aside from nuclear war, it would be hard to imagine a faster or more thorough way to destroy land, soils, rivers, aquifers, rural communities, and lives. We have better alternatives, however, that go back to ideas of peasant farmers in China, J. Albert Howard, Lady Eve Balfour, J. Russell Smith, tree croppers, forest farmers, soil conservers, permaculturalists, and people who just pay attention.

In the recent history of alternatives to industrial agriculture, Jackson is reminiscent of Louis Bromfield, the founder of Malabar Farm in the late 1930s.[2] Bromfield, too, challenged orthodoxy and farm practices through demonstrations at Malabar, weekly tours, and five books. Bromfield was one of the mid-twentieth-century group that aimed to improve American farming that included Aldo Leopold, J. Russell Smith, Paul Sears, E. H. Faulkner, and Russell Lord. They failed, however, to deflect the juggernaut of capital, farm technology, and hubris that was industrializing agriculture.

Farms became larger by absorbing their less profitable neighbors. They also became more specialized and capital intensive and so more vulnerable to the slightest twitch in the economy, with results visible throughout rural America. With notable exceptions, such as Joel Salatin, the countryside has been emptied of diversified small family farms and much of its loveliness. Most small towns are on life support, and the reservoir of skills once learned on diversified small farms has all but disappeared. The reasons are many but boil down to national policies that keep the price of food low in large measure by ignoring the ecological and human costs of agribusiness. Corporate farmers, as Marty Strange once said, became more adept at farming the tax code than growing crops.[3]

Jackson, however, takes the story to a deeper level. The problems *in* agriculture, as he puts it, are less important than the problem *of* agriculture. Soil erosion, destruction of biological diversity, and dying small towns are simply the remorseless working out of the rules of a subsystem that conflict with those of the four-billion-year-old ecosphere of which it is a part. Uncorrected, it is a lethal miscalibration. Consequences, in other words, have causes rooted in the ideas—however mistaken, hairbrained, or noble—for which we live and die. For better or worse, those same ideas form the paradigms and models for agriculture, economies, global supply chains, the banking industry, political systems, public institutions, consumer-oriented societies, and systems of technologies with effects that no one understands.

Globalization, for example, designed for maximum economic growth and minimal public oversight, has performed miracles on one hand while creating pandemics, massive inequality, deforestation, soil erosion, pollution, acidification, widespread obesity *and* starvation, oceanic dead zones, and climate chaos on the other. The latter are no more accidental than the former. Given the rules of the system, time, and circumstance, all are predictable outcomes.

Of the many things that can be said of Wes Jackson, the most important have to do with his role as a master teacher who taught whoever was listening to think more imaginatively and courageously about our connections to soil, culture, time, and how we got to our present state. As I write, a pandemic is raging across the world, and it, too, is the logical working out of the rules of the larger system overriding those of the smaller system. In my home state of Ohio, many living in rural communities are dying early from opioid addiction and others from despair.[4] That, too, is a systemic problem in which the rules give priority to profits for a few at a heavy cost to many.

Solutions, when they arise, are not likely to come from current paradigms that give priority to short-term profit, quick results, and domination of natural systems that we do not understand. Europeans arrived in what they thought was a new world but with habits and minds formed in the old. In the words of one historian, they came as "hustlers" aiming to build empires of one sort or another. We live in the ruins of those ambitions, not necessarily the ones that failed but those that succeeded too well. One hopes that their descendants will learn how to be ecologically competent "homecomers," in Jackson's word: rooted in place; adept at neighborliness, repair, frugality, and humility; and practitioners of democracy worthy of the name. Jackson would hope that they will also be good storytellers around future campfires.

*Introduction:
In the Barn*

For regulars at Prairie Festival, The Land Institute's (TLI's) "Big Barn" is a kind of secular church to which we return for our annual perennial inspiration. Prairie Fest might best be described as a tent revival meeting fused with a graduate school seminar—updates on plant breeding, a lot of political/economic analysis, some cultural criticism, and along the way a bit of spiritual uplift.

No one is selling salvation in conventional religious terms, but the conversations that swirl in and around the barn on those September weekends deal honestly with the problems we have to confront if our species is going to be saved—or, at least, if our species can demonstrate that it deserves to be saved. These are the conversations that bolster the commitment of Prairie Fest veterans to continue struggling to help create a decent future.

To a first-timer, it might seem to be an odd gathering—hundreds of people from around the country, and a few from across the globe, jockeying to sit on folding chairs on a dirt floor, whatever the weather, in a barn that's open on two sides on the outskirts of Salina, Kansas. But that dirt under our feet—that is, the prairie soil and the struggle to keep it healthy and in place—is at the heart of the project. As TLI cofounder and longtime president Wes Jackson puts it, "The only way to save our souls is to save our soils." If we cannot manage to care for a resource that is essential if we are to feed ourselves, what hope do we have?

When Jackson lets rip with an aphorism like that—pithy and memorable, not just for fun but conveying a deep truth—those in the barn feel both the excitement about the work going forward on the property and the fear of a frightening future that makes TLI's work on perennial polycultures (as Jackson is fond of saying, "More on that later") so important.

For the better part of four decades, until his retirement from the job of TLI president in 2016, Jackson's talk would close the festival on Sunday at noon. His talks ranged widely over matters ecological and social, but whatever the particular subject of any talk, the long-term goal was connecting with what he loved to call a "consecrated constituency," the people who have supported the work and spirit of the organization and embraced an ecological worldview.

A good example was his talk at the 2011 Prairie Festival, titled "The Next Synthesis," which was based on a paper he had written with two TLI staff members, ecologist Tim

Crews and sorghum breeder Stan Cox. The paper had a typically ambitious goal. After outlining three previous breakthroughs in biology (Darwin's theory of evolution by natural selection in 1859, Dobzhansky's bridge between natural history and genetics in 1937, and Watson and Crick's elucidation of the genetic code in 1953), Jackson made clear his intentions: "a fourth synthesis on a global scale—not for its own sake, but to make it possible, for the first time in history, to provide humanity with a sustainable foundation in grain agriculture. It has the added potential to turn our industrial minds toward an ecological worldview precisely because it features material and energy limits which paradoxically become a source of new possibilities."[1]

That's somewhat audacious—placing oneself and colleagues in a lineup with key figures in modern biology, seeking to reinvent agriculture to give our species a shot at long-term survival, and advocating a foundational shift in how humans should understand their place on the planet. But after decades of work on Natural Systems Agriculture (NSA—those perennial polycultures), there is justification for boldness on the part of Jackson and the scientists at TLI, where he always encouraged a swing-for-the-fences approach to cracking tough problems.

But in listening closely to Jackson, it becomes clear that the boldness lives alongside a deep humility. In lectures and writings, he advocates passionately for his views but rarely proclaims "the truth." Aware of human limitations, he highlights crucial questions and formulates answers that invite more discussion. He will sometimes preface a remark with "I'll make a bold statement here, which I'm prepared to talk about for the rest of my life." We face ecological and social problems that have no easy answers—perhaps no real answers at all if we allow our thinking to be trapped within existing political and economic systems. No one has grounds for being cocky.

Jackson's concern about the dangers of getting stuck in a narrow approach to problem-solving was evident after the formal part of the 2011 festival ended, as he made the rounds of the lunch tables set up in his backyard overlooking the Smoky Hill River, where TLI staff and friends every year relax after the hundreds of visitors have left. It was only my second visit to TLI, and I was staying at Jackson's heels, collecting material for a magazine article.

Leaning into one group, where the topic turned to the problem of arrogance and certainty, Jackson suggested that an important first step to solving big problems such as unsustainable agricultural systems is to recognize that sometimes "we've got to give up on what we know."

If there were one sign he could hang above everyone's desk, Jackson said, it would be this daily affirmation: "This day I will do everything I can to fight the problem of reassertion." Reasserting what we think we know for sure, over and over again, is trouble, especially in the sciences, he said.

Jackson—raised on a farm and trained as a botanist and geneticist—isn't antiscience, of course. He is trying to bolster, not undermine, faith in science by encouraging practitioners not to get too comfortable. In agriculture, such groupthink has led researchers to assume that the so-called Green Revolution emphasis on chemicals, fossil fuels, and advanced technology is the only way to maintain high yields. Research on initiatives such as perennial grains grown in polycultures, Jackson argues, may well reveal the conventional wisdom to be conventional foolhardiness.

With the health of our soils and our own bodies at stake, Jackson believes that we cannot afford to assume that old approaches are adequate to cope with current crises, let alone with what is coming. Modern humans seem to like to resolve conundrums as easily as possible, and institutions reward researchers who appear to "solve" tough problems within existing systems. Such research may be right in some sense, but it is irrelevant if the real problem is at the level of the whole system; solving individual problems within a system that cannot be sustained is, at best, a temporary patch. Finding a slightly less toxic poison to kill weeds keeps us trapped within farming-by-poisons.

TLI—and science more generally—won't provide all the answers. Jackson recognizes that serious changes are necessary in social systems as well. He takes every chance to challenge modern systems' love of abstraction. In contemporary financial capitalism, he said, countless decisions about money are based on abstraction, not on the material reality of economics rooted in ecosystems. Leaning into the table a bit more, he said, "Milton and Blake both were on to that, that the demonic is the abstraction without the particular."

The particular is where we test the accuracy of our abstractions, and science will help resolve the ecological crises only when it remains rooted in that particularity. Farmers work the land in a specific place within a specific ecosystem, where they must attend to the uniqueness of place, Jackson says. An idea such as perennial polycultures is valuable not as a monolithic answer in the abstract but as an idea tested in specific places, whether that be Kernza® fields in Minnesota or rice fields in China.[2] Jackson was never out to make TLI the center of sustainable agriculture but instead wants to see the ideas developed in as many places as it is sensible to do so.

Jackson also cautions that our specific places should be understood as part of larger systems, and that project is both theoretical and experiential, analytical and emotional. To experience our place in that larger living world, we sometimes have to step outside science. Jackson is a very dialectical kind of person, although not the kind of person likely to use the term "dialectical" in conversation.

Around the tables, even though the TLI staff are at the end of an exhausting three days of hosting their major event of the year, Jackson was still revved up.

"We know the earth revolves around the sun, but our daily experience is of standing on ground that doesn't move," he told the folks at a third table. To correct that, we should take the time to feel the earth move. Jackson was off and running:

I have actually felt the earth turn. I can tell you how to do that. I've gone out there and laid down on the hill when the moon is full, and if you will look when the moon is coming up in the east and the sun is setting in the west—you've got to live in Kansas to do that, or Nebraska, someplace flat—you can actually feel the earth turn. Do that sometime. It's a great moment. You've got to do that extra exercise to experience reality. Otherwise we live with the illusion.

After a brief pause and a quick smile, Jackson added, "Which is fun enough."

Jackson took a moment to delight in both his memory of the experience and the smiles on the faces of the people at the table. Then he said, "I suppose that in order to experience reality, you have to be bit of a mystic," and moved on to the next table.

Hands-on Mysticism

If Wes Jackson is a mystic, he is the hands-on type. Jackson loves working with his hands and takes pride in showing people the structures at TLI that he helped build, along with the details of how much was built with scrap lumber or old railroad ties. He loves to tell the story of going to the county office to get a building permit for the original house on the property. The clerk asked to see his plans. Grabbing a tablet on the counter, Jackson sketched the outline of the house freehand, drew in a couple of doors, and submitted the "plans," which were promptly approved.

Jackson also loves the prairie, and he'll grab any opportunity to jump in the car to give visitors a tour, usually ending with a stop to see the handful of cows he raises that are grazing on land near his home. A drive through the Kansas countryside with him makes it clear that Jackson never stops thinking about the land and never tires of talking to people. Specific places remind him of events in his life (he grew up on a small farm near Topeka) or the state's history (this is "Bleeding Kansas," in some sense the place the Civil War started), and all these observations kick off new lines of inquiry. Passing through Wellsville, just south of Lawrence, Jackson remembers that his mother's first cousin, Dubreeze E. Pearson, who had dedicated her life to journalism at the *Wellsville Globe*, had made a substantial contribution to build the town's public library, and he was interested in seeing it. The librarian on duty was delighted to show him Dubreeze's desk, which stood as a memorial to her work.

He reacts to the beauty of the Flint Hills region as if it's the first time he has ever seen the rolling ranchland, much of which has never been broken by a plow. "Just look at that," he says often. Stop in a small town and Jackson will seek people out to chat, as if he's lived there his whole life. For Jackson, the world presents endless opportunities to get to know people, other living creatures, and the nonliving components of landscapes. That same world is also an endless source of conundrums to puzzle over. Both the interactions and the thinking are an endless source of pleasure for him.

But Jackson also worries as much as anyone I have ever met about the future of the species and the land on which we live. Because he genuinely likes people and creatures, he spends a lot of time trying to understand the impediments to ecological sustainability and social justice that we humans have created. Jackson has lots of questions on his mind, lots of problems he would like to help solve. That thinking is a source of endless concern and a certain amount of existential pain.

These questions do not have easy answers, and the problems do not have easy solutions. In fact, Jackson gravitates toward problems that have no solutions, if by "solutions" we mean coming up with ways to continue business as usual in a high-energy world with lots of gadgets. His life's work has been to face down the intractable problems precisely so that any solutions we claim to devise might actually help rather than merely papering over what we cannot bear to face honestly. Jackson believes we can stare down these questions and problems while continuing to live and love, out of his conviction that facing these harsh realities will make possible "a more joyful participation in the Creation."

What does this joyful participation look like? With an elegant simplicity, Jackson calls it the "mill around" philosophy of life, expressed in the form of a question: "How do we mill around, amuse ourselves, and live cheaply until we die?"

We need to keep busy, we want to stay upbeat and at least occasionally have some fun, and we have to learn to do that without burning up all our options for the future. "Done right," Jackson says, "think of what a joyful journey this could be." Call it a mill-around mysticism, sometimes expressed through the Christian and Jewish stories he grew up with in the Methodist church, but without dogma. Jackson does not hesitate to talk of the Creation without asserting that there is a Creator. "There's Methodism in my madness," he is fond of saying, as a way of honoring his family history and recognizing the power of the stories of scripture.

To mill around sustainably, Jackson argues, we need to change the way that most contemporary humans see and understand the world around them and the world within them. We all need to adopt an ecological worldview—to see the Earth as alive rather than a mine from which we extract resources and a dump for our wastes.

"We've got to get some things straight real quick," Jackson says these days. There's not all that much time, given that we live in a world still obsessed with economic growth, with people increasingly tethered to screens for on-demand amusement, racing around at an ever more frenetic pace that burns more fossil fuel that closes off more options that leaves us trapped in the industrial worldview.

Prophetic Stories

It is that different way of seeing that world that first drew me to the work of Wes Jackson. When a friend recommended—

more accurately, demanded, given my friend's sense of how little time we have to get things straight—that I read Jackson's first two books in the late 1980s, I was skeptical. As a first-year graduate student focused on political and mass media issues, I wasn't sure what I would get out of a plant geneticist's writing about agriculture. That recommendation turned out to be a turning point in my life. I did not change my field of study, but I realized that any investigation into human societies was at best irrelevant and at worst dangerous if it did not take seriously a deep history of our species and the planet—that is, everyone in every discipline should be trying to better understand how humans ended up in this position of dominance that is undermining our future.

That was for me the beginning of decades of playing intellectual catch-up, trying to understand a history that had never been taught in any of the classrooms I had sat (or slept) in. For three decades I read everything Jackson published—articles, books, even fundraising letters. Jackson's annual pitches for TLI were the only such letters I read in their entirety and enjoyed.

Just as crucial to that ongoing education were the tapes from Prairie Festival, which Jackson loves to describe as "an intellectual hootenanny." Long before I ever attended the annual event, I listened to tapes of Jackson's festival-closing talk (yes, cassette tapes at first, later CDs, now online), which my friend and I copied and shared widely (a confession to multiple copyright violations over the years).

It was Jackson's oral storytelling that really put the hook in me. Woven into his analysis are characters and experiences—from his childhood days on a Kansas truck farm through university training with brilliant and quirky professors, being an early shaper of the field of environmental studies, cofounding TLI, and pushing forward innovative research on perennial polycultures.

Listening to Jackson's voice on tape led me to think of him as a kind of prairie prophet—not someone who claims to see the future but a critic speaking in the prophetic voice, invoking the prophetic tradition. This term is often used as a synonym for speaking out against, and taking action to stop, injustice and oppression. But the truly prophetic analysis offers not only a critique of the obvious targets in positions of illegitimate authority (such as kings, queens, corporate CEOs, and investment bankers) but a moral challenge to everyone (including ourselves and our loved ones) to face difficult truths and acknowledge our own complicity. Jackson routinely reminds audiences that all of us living in an affluent society are caught up, at some level, in the structural immorality of the system. Quoting Wallace Stegner, he emphasizes how easily we can become the prisoner of "things that once possessed could not be done without."[3] Borrowing a phrase from George Orwell's critique of left-wing parties in Britain, he worries that environmentalists are as hooked on dense energy as anyone and therefore "make it their business to fight against something which they do not really wish to destroy."[4]

Jackson engages in a loving but blunt criticism that evokes the prophets of Hebrew scripture. In his analysis of these critics, scholar and activist Rabbi Abraham Joshua Heschel concluded, "Above all, the prophets remind us of the moral state of a people: *Few are guilty, but all are responsible.* If we admit that the individual is in some measure conditioned or affected by the spirit of society, an individual's crime discloses society's corruption" (emphasis added).[5]

Taking one's place in that prophetic tradition does not require self-aggrandizement—any farm boy can take a shot at it. Remember that those Hebrew prophets did not see themselves as having special status but rather were ordinary people. When the king's priest confronted Amos (one of the so-called minor prophets) for naming the injustice of his day, Amazi'ah called Amos a "seer" and commanded him to pack his bags, head to Judah, and "never again prophesy at Bethel, for it is the king's sanctuary, and it is a temple of the kingdom." Amos rejected the label:

> Then Amos answered Amazi'ah, "I am no prophet, nor a prophet's son; but I am a herdsman, and a dresser of sycamore trees, and the Lord took me from following the flock, and the Lord said to me, "Go, prophesy to my people Israel."
> (Amos 7:14–15)

Jackson may not be a herdsman or dresser of sycamore trees—he has never been a full-time rancher or farmer—but he remains a person at home in ranch and farm country, as comfortable pulling up in his pickup to ask a neighbor about the health of cattle as he is delivering a lecture in a university hall. In those lectures, audiences will hear him quote Dante and the Hebrew Bible as he explains ecosystem functioning and the important differences in the root architectures of various plants. The force of the "prophecy" he delivers—that call to save our soils and our souls—emerges out of the sometimes jarring juxtaposition of his sources and metaphors. Whenever you think you know where Jackson is coming from, he's likely to come at you from a new angle you hadn't considered, and he may not have considered until that moment.

Restless and Relentless

Those varied angles of vision are not surprising if one knows a bit of Jackson's biography. "Born on the end of a hoe handle" in 1936, he got his primary education in a two-room rural school and in his family's fields before leaving for Kansas Wesleyan University (KWU) in Salina to play football, run track, and study biology, in roughly that order of enthusiasm, graduating in 1958. After earning an MA in botany at the University of Kansas in 1960, he taught biology and coached at Olathe High School outside Kansas City and then at KWU. After completing a PhD in genetics at North Carolina State University in 1967, he returned to KWU to teach before heading out in 1971 to California State University, Sacramento, which hired him to start an environmental studies department. After that project was up and running, he took a one-year leave of absence to return to Kansas, eager to buy land and homestead. That leave stretched to two years, and the Jacksons never returned to Sacramento. The small piece of property that he and his family were living on outside Salina became TLI, launched as an alternative school in 1976.[6] Jackson admits that he left the security of a tenured job as a full professor with some trepidation, but from that modest start, TLI grew to be an internationally recognized center for sustainable agriculture research. At the age of eighty, Jackson retired, sort of. Turning over the job of running TLI to a new president, he helped launch a new ecological education project dubbed Ecosphere Studies[7] and turned more of his attention to writing, including a book of stories from his life titled *Hogs Are Up: Stories of the Land, with Digressions.*

How did a Kansas farm boy accomplish all that? More importantly, what led that farm boy to leave a conventionally successful academic career path without much money in the bank and sketchy ideas about how TLI would survive?

Jackson would be the first to point out that there's a fair amount of dumb luck involved, but just as much can be explained by his restless and relentless mind, which keeps pushing him to find new challenges. When Jackson is poking at an audience, asking them to think in new ways, he is really just reporting back to the group about how he pokes at himself. While lots of people are diligent and curious, Jackson's near constant state of intellectual agitation is unusual. In 2019, after Jackson slipped on an icy sidewalk and suffered a concussion, the doctor advised him not to look at screens, read for long periods, or do much intellectual work for a couple of weeks. Jackson was able to handle the first two requests, but the third was impossible. "How am I supposed to stop thinking about things?" he asked. "Am I supposed to be Buddha?"

When Jackson sinks his teeth into a question, he doesn't let up until he has figured out an answer that he is satisfied with, at least for the moment. And once he has an answer—at least a provisional one—he wants to move on to the next challenge. He doesn't mind the recognition that his work has brought him—a Pew Conservation Scholar in 1990, a MacArthur Fellow "genius grant" in 1992, the Right Livelihood Award in 2000, and one of the Smithsonian's "35 Who Made a Difference" in 2005—but he isn't interested in coasting on his accolades or his considerable accomplishments. Jackson likes being right as much as the next person, but what really gets him excited is a difficult question that brings him to the edge of a new insight. He seems genuinely happy when he is talking with someone who makes him rethink what he thought he knew.

One journalist's description of him as "a large man with the metabolism of a hummingbird"[8] pretty much captures the experience of being with Jackson when he has a new idea. Friends and colleagues have all had the experience of him bursting into a room or starting a phone call with a new idea, so excited that he forgets about customary greetings. Once the idea has been unleashed, he will double back and say, "And, by the way, how art thou this morning, Brother?"

There is a pattern in Jackson's rapid-fire thinking-aloud process, perhaps best described by journalist Adam Gopnik's definition of "the historical imagination" as "the ability to see small and think big."[9] Jackson naturally toggles between the two, paying attention to the details of the world and then stepping back to see the patterns of which the small things are a part. That's the dialectic between the particular and the abstraction that helps us steer clear of the demonic. Jackson's early work in botany was motivated by a desire to know the details about specific plants, and that same interest in particulars took him into genetics. But he quickly expanded into ecology and environmental studies, seeking to understand the patterns that shape the trajectory of human history as part of Earth's history.

Getting a Handle on Complexity

Jackson has advanced training in botany and genetics, but his most important contributions are not in those specialized fields. His restless mind leads him all over the intellectual map, which comes with advantages and disadvantages. Focusing on one discipline, or these days on one subfield within a discipline, allows one to speak authoritatively on questions in that field. But that kind of "seeing small" can undermine the ability to make sense of larger patterns. "Thinking big," however, can lead one to get sloppy on the small stuff in the quest to identify those patterns. Finding the right balance is a challenge not only in scholarly work but in human life generally. Our attempts to categorize and explain a complex world, which allow us to get through the day successfully, inevitably simplify that world in ways that can lead us astray. Such is the human condition, and the frustrations of the human condition are magnified by the turf battles and status contests that arise in academic life.

Too often, when a generalist makes claims, the specialists are quick to point out what the generalist has misunderstood or overlooked. Such critique is important when it corrects errors, but at times the critic's goal is simply to eliminate a challenge from outside the discipline in order to protect reputations. Specialists often point out that a generalist seems unaware of the "latest" developments in a scholarly discipline, as if the alleged deficiency automatically negates any insight.

My tone might suggest that my sympathies lie with the generalists, which is partially true. In thirty years of academic life, I saw how the system of rewards and punishments favors the specialist, to the detriment of a broader intellectual life. But it's impossible not to acknowledge the accomplishments of specialists, without whom the generalists would be adrift.

My sympathies really lie with those committed to creating knowledge that can help us understand the human condition and provide us with tools of finding our way to just and sustainable societies. Jackson is one of the most important thinkers on my list, in part because he makes no claim to special status. His insider experience in academia taught him the value of rigor, and his outsider status since leaving the academy has freed him up to range widely across disciplines.

A Collaborative Effort

This book organizes some of Jackson's key insights from his lifelong efforts to see small and think big. Starting with the most basic ways to understand life on this planet, we'll hit the highlights of his intellectual journey, presenting the worldview that Jackson has been constructing for many decades. For longtime readers of Jackson, many of the phrases will be familiar but may be organized in a new way that can spark new thinking. For those unfamiliar with his work, the book can serve as an introduction to the worldview that Jackson has built from a lifetime of experience, drawing on a range of disciplines.

Woven throughout the ideas will be lessons he learned on the farm alongside formal scientific conclusions; history and theology, in both cases understood as stories that bring to life human wisdom rather than ironclad empirical claims about the past; and the influence of what Jackson calls the "indispensable friendships" that have influenced him, with people ranging from the acclaimed writer Wendell Berry to Leland Lorenzen, a quirky neighbor who was a twentieth-century Kansan version of Thoreau.

When I retired from teaching at the University of Texas at Austin in 2018, I eagerly threw myself into this project. Writing a book about, and with, Jackson felt like coming full circle. Reading, especially in our formative years, can be a search for the like-minded before we know enough about our own minds to understand what we are looking for. That was certainly true of me when I first picked up Jackson's books back in 1988. Long before I could articulate why, I sensed that this reading was transformative for me, part of shaping the worldview that would serve me well through my teaching career, political organizing projects, and own writing life.

While working on this project, Jackson and I maintained an extended conversation in person, on the phone, and in writing. In that process, I found him to be incredibly generous, not only in sharing his thoughts but in letting me see his thought process unfold in real time—a window into the workings of that restless and relentless mind. Though Jackson has never been an artist in traditional terms, those conversations reminded me of the creative aspects of scientific and analytical thinking.

Though I offered my own observations and ideas in those conversations, I am hesitant to claim anything in this book as "mine," given the degree to which Jackson's work has influenced me for three decades. My contributions to the collaboration came mainly in reacting to Jackson's creativity, building on his insights.

Yet while this is a book to convey the core of Jackson's work, it reflects the way I came to understand that work and, in the process, understand myself—our attempts to understand the world emerge from our hunger to know ourselves. Let me state the obvious: My account of Jackson's ideas is inevitably both skewed and incomplete. The focus is on those ideas that I found, and still find, most powerful, organized in a flow that I hope is elegant enough to do justice to his projects. In places, I add my own observations to round out Jackson's work, even though it's futile to try to draw a bright line between his ideas and mine. Chapter 8, an essay about "the creaturely worldview," is a recent essay that Jackson and I wrote together.

In the end, this book cannot be anything more than a synopsis that should lead people to the written and audio/video record of Jackson's work. There, in his own words, readers will find the challenge and the pleasure of engaging Wes Jackson's restless and relentless mind.

1

The Energy of Life

Wes Jackson spent much of his youth, along with his parents and five siblings, helping to grow and sell produce from their farm family as well as working on other Kansas farms and a South Dakota ranch. He worked in a foundry as a welder to pay for college, which led to a teaching career—all before settling in at TLI for more than four decades.

But whatever the particular job Jackson has held at any point in his life, he's always been at his core a storyteller, which requires being a keen observer as well as a fast talker. Many of those stories are drawn from his life, populated with a wide range of characters from the farm, classrooms, laboratories, and political movements. But just as important to Jackson is telling the story of life on this planet, of how the incredible diversity of living things on earth came to be.

Humans have always told stories about our beginnings—every culture has origin myths—but Jackson points out that modern science now gives us a story that is based on "a verifiable cosmology." Physicists haven't worked out all the details—there are, after all, those pesky questions about dark matter and dark energy—but we have considerable understanding of the origins of the universe and its trajectory forward from the big bang. Chemists and biologists haven't worked out all the details, either, but we also have a way to grasp the origins of life that helps us understand "the journey from minerals to cells," as Jackson likes to call the process.

But like any good scientist—and any curious layperson—Jackson knows that it's productive to rethink even the most basic questions now and then. That's why in his eighties, Jackson is still pondering the simple question: What is life? As for so many simple questions, the answer turns out not to be so simple.

Though trained as a geneticist, Jackson doesn't start the inquiry about life with DNA, nor does his botany training and farm background lead him to focus on organisms. Jackson says it's misguided even to focus on the concept of the "biosphere." Instead, he begins at the planetary level—not only the biosphere but also the atmosphere, lithosphere, and hydrosphere. Because all are essential parts of a living whole, Jackson takes the foundational unit of analysis to be the *ecosphere,* the proper "boundary of causation" (taking into consideration all the forces and materials that create our world). At the next level down in the hierarchy of structure, *ecosystems* become the primary focus for human investigations as the "boundary of consideration" (the scope of what humans can reasonably pay attention to).[1] To the degree possible, the boundaries of causation and consideration should line up so that when we make knowledge claims, we are taking into consideration all the relevant causal vectors.[2]

In other words, everything on Earth really is connected, but the human mind cannot connect all the dots on all places on the planet. We can, however, pay attention to ecosystems and study not just the components of an ecosystem but ecosystems themselves, as whole entities with properties of their own. That is what the late University of California, Berkeley professor Arnold Schultz called "Ecosystemology,"[3] an approach that Jackson embraced even though the term never caught on in ecology. Instead of thinking about organisms in their environments, which leads us away from the connections, this approach starts with ecosystems, which include organisms.

That leads to the late ecologist J. Stan Rowe, a Canadian contemporary of Schultz whose ideas had a strong influence on Jackson's thinking. Rowe argued that instead of "organism = life" (the idea that life is a property of an organism) we should revive the more ancient idea that "Earth = Life" (life is a property of the ecosphere). This challenges us to ponder the way we divvy up the world between the living (such as plants and animals) and the dead (such as water and rocks). To challenge the living/dead dualism, Rowe said that if we could shrink ourselves and travel inside a cell, it might appear that there are some moving/living parts and other static/nonliving parts. But from the outside, the whole cell is seen as living, the result of the participation of all components. Here's how Rowe explains it:

> Analogous to the outside/inside view of Earth is examination of a cell under a microscope, followed by an imaginary view from within. Looked at from the outside the cell is seen as a unit whose parts are the watery cytoplasm, the vacuoles, inclusions such as starch grains, the nucleus and various other organelles. Seeing the whole, the viewer accepts at once that all the constituents are related components of a living cell. Now suppose the viewer is reduced to micro-size and placed inside the same cell with a pair of binoculars. Looking out and around s/he will apparently see the same cell components but now as separate things. The slow flowing cytoplasm, the vacuoles and starch grains, will appear "dead" while the more active, dividing organelles will be identified as "alive." From the outside all the components participate in and express the life of the cell. From the inside only certain parts appear to be animated. Just so, people immersed in Earth's surface—deep-air animals—have misclassified most of what lies around them as "dead."[4]

To challenge the living/dead dichotomy is not to suggest that rocks have consciousness or that a piece of iron ore has the same moral status as humans. Rather, it reminds us to be careful about saying that any part of the ecosphere is "nothing but." For example, people too often will consider soil "nothing but dirt," Jackson says, obscuring the complexity of life in that soil. Not only is the soil "alive" in the sense that it is teeming with microorganisms but the "living" components of that soil could not exist without the "dead" parts. Jackson points out that when we carve up the world into these categories of living and dead, "we find it easy to play fast and loose with the parts." We treat the planet like a mine from which we extract the resources we want and a dump for wastes we don't. The mania for that extraction is less surprising when we understand life to be . . .

The Scramble for Energy-Rich Carbon

OK, so back to the question: What is life? It's fine to set the question in a larger ecospheric context, but in everyday life we still talk about some things being alive and some not, and the difference matters in how we treat things. Biologists constructing a checklist of the features of living things typically include the capacity to grow, metabolize, regulate the internal environment, respond to stimuli, adapt to their environment, reproduce, and evolve. One often-quoted definition of life is "a self-sustaining chemical system capable of Darwinian evolution."[5] But for all the attempts, there is still no single, universally accepted definition for life. Biologists would all agree that bacteria are alive but argue about whether a virus is alive.[6]

If one goes on too long at this level of detail, Jackson says, "people start heading for the exits." When offering his answer, Jackson isn't trying to settle a complex scientific question but instead wants to focus our attention on one crucial feature of living things: Life is the scramble for energy-rich carbon.

Life on this planet is carbon-based, and capturing carbon gives us things we need (the energy in food) and things we like (the energy in fossil fuels that we burn to do lots of work for us). That definition puts what Jackson calls "the carbon imperative" front and center, reminding us of our animal nature. We humans are like all other living things, but we have gotten exceptionally good at using not only the carbon produced by contemporary sunlight but the carbon produced by ancient sunlight that is trapped below ground. Given the variation in culture and individuals over the course of history, "human nature" is notoriously difficult to define with precision, but at the core is our "carbon nature." Bill Vitek, a philosopher and one of Jackson's longtime collaborators, suggests that we would be better served using the term "human-carbon nature" to remind us of our carbon-seeking proclivities.

Jackson says he began pondering this way of defining life in the early 1970s, when he was on the faculty at California State University, Sacramento. One of his colleagues in creating the new Environmental Studies Department there, Chuck Washburn, was a mechanical engineering professor who brought with him a focus on energy and energy sources, subjects of considerable interest and debate today. Ask Jackson where an idea of his came from, and he's likely to tell a story about how someone in his life helped him understand the world in a new way, and Washburn is one of his friends who pops up in a lot of those stories.

But Jackson also puts a distinctive twist on these ideas, and in this case the focus on the carbon imperative and humans' success in extracting energy from the planet leads to him talking about where the carbon that changed us—and the world—has come from, what he calls the . . .

Five Carbon Pools

For most of our evolutionary history, humans met their carbon needs through gathering and hunting, activities that produced little surplus and could do limited damage to other species and ecosystems. All of that changed with agriculture, when humans figured out how to store energy in grains. We'll return to the importance of the domestication of plants and animals in the next chapter, but for now we will stay focused on carbon.

Jackson suggests that one way to understand the past ten thousand years of human history is to identify where humans have found all that carbon, the five carbon pools. "We got very good," he says, "at skating from one carbon pool to the next."

The first pool of carbon that humans began using—and wasting in dangerous ways—was the carbon in the soil that was exploited through agriculture, what Jackson refers to as "the young pulverized coal of the soil," a phrase he borrowed from the energy-efficiency guru Amory Lovins.

The second pool of carbon came from the world's forests. Agriculture led to more people, living in increasingly larger urban areas, and so more trees were used in construction of various kinds and burned to keep us warm. But more crucially, forests were felled to provide the carbon needed to smelt ore in the Bronze and Iron Ages, which gave rise to what we call civilizations.

The third pool was coal, a revolutionary fuel when it could be burned to generate steam in the engines that drove the Industrial Revolution. This carbon could replace human and animal labor, doing incredible amounts of work for us. Not far behind were the fourth and fifth pools of carbon, oil and natural gas, which made possible the modern high-energy/high-technology world that many of the world's nearly eight billion people take for granted and most others yearn to be part of.

In some times and places, some people—either as part of a social unit or as individuals—have successfully resisted the maximal exploitation of all the carbon that their level of technology makes available to them. But the temptations of dense energy are powerful for humans for the simple reason that we are organic creatures and not angels. Human nature is carbon nature. That does not mean we cannot modify our behavior in light of new knowledge about the negative effects of excessive carbon-seeking, but our chances of successful modification are increased if we come to terms with that reality—humans tend to use all the carbon they can get.

This recent history of intensified carbon-seeking can be described as the "hole-poking" phase of human history. For ten thousand years, Jackson says, people have been poking holes in the ground to get at as much of that energy-rich carbon as possible. But we didn't evolve as hole pokers, which is why it is not surprising that we are always messing it up, Jackson says. We are best understood today as being . . .

A Species out of Context

Perhaps the most puzzling paradox of our moment in history is that we humans are smart, but for all our intelligence we routinely engage in behavior that likely will make a large-scale human presence on the planet impossible in the not-so-distant future. That's not very smart but is easier to understand when we recognize that we are "a species out of context," which is the Jackson aphorism that has had the most long-lasting effect on my thinking.

Let's start with a reminder of the timeline: Before the invention of agriculture and surpluses of grain starting 10,000 to 12,000 years ago, human beings were all foragers. For the

2.5 million years of the genus *Homo* and 200,000 years of *Homo sapiens*, everyone lived as gatherers and hunters. We evolved in band-level societies of 15 to 50 people, with connections that probably didn't exceed 150 and living arrangements that generated relatively egalitarian social relations with little or no hierarchy. Humans gathered plant foods and hunted, typically ranging over a territory with frequent movement.

The surplus-and-hierarchy system predated agriculture in a few resource-rich places, but the domestication of plants and animals spread that dynamic across the globe.[7] Especially where grain crops dominated, the ability to create and store surpluses generated hierarchies that have produced profound social inequality.[8] There were limited forms of settled community life in some places prior to agriculture, but it was only after people started farming that we got cities, empires, nation-states—all forms of social organization based on hierarchy. Not all people in all places developed on the same trajectory—differences in climate, geography, and environmental conditions have shaped human societies differently[9]—and in some places people rejected such organization. But eventually the size and scale of human communities in most places were far beyond anything we evolved in.

Because our evolutionary history shaped us for a very different way of living in much smaller social organizations, it's not surprising how hard we have to struggle—and how often we fail—to make our out-of-context social systems work to everyone's benefit. Most, if not all, of our current social and ecological problems are a result of humans trying to live in political and economic contexts for which we are not adapted in evolutionary terms.

Of course, Jackson is not the first person to highlight this human trajectory—it is the basis of many an anthropology and history course. But in assigning to that history the observation that we are a species out of context, Jackson captures the problem we face in ways that both challenge and comfort us. Saying that "we" are a species out of context doesn't mean everyone in the history of the species is equally culpable for ecological degradation (more on that in Chapter 7), but it highlights that we are all part of one species. Humans respond to not only social but, like all other organisms, biological imperatives. We all have a lot of work to do if human history is to have a future—that is the challenge, to be the first species that self-consciously practices limits (more on that in Chapter 7 as well).

2

Agriculture

When humans moved from a gathering-hunting economy to an agricultural economy, human social organization changed, and so did our relationship to the larger living world. Instead of relying on ecosystems managed by natural processes, humans began intervening to select certain plants for cultivation and mark others for elimination, to alter landscapes, to manage the flow of water on a large scale. We also began domesticating animals to supply us with food and other materials. For Jackson, this is the beginning of our break with nature, the point at which humans moved from living off the land to actively manipulating the land for a living. We took charge of the fate of other living things in a dramatically new way.

The problem with all that, Jackson says, is that we are not very good at it.

This challenges the narrative most of us have learned about the human journey from the so-called primitive to the civilized, in which agriculture marks the beginning of the rise of humans to new levels of creativity and achievement. But if we are not careful, Jackson warns, agriculture may end up being the fall from grace from which we never recover, which means we need to recognize . . .

The Problem of Agriculture

For someone born on a farm, who loves talking about farming with farmers and has worked most of his life searching for better ways to farm, it seems odd to hear Jackson say that rather than focusing on problems *in* agriculture, we must deal with the problem *of* agriculture. To be clear, Jackson is not blaming farmers but rather focusing on the consequences of farming, not just in the era of industrial agriculture but going back to the earliest domestication of plants and animals.

Agriculture arose independently in at least eleven different places, the first likely being the growing of wheat in what today is Iran. Though there is variation in the ecological consequences of different styles of horticulture and agriculture, the annual grain agriculture that gave rise to early civilizations—crops that must be planted every year and usually require disturbing the soil—has consistently led to soil degradation and erosion beyond replacement levels. That's the problem *of* agriculture. And though there is variation in the social structures of agricultural societies, the political and economic hierarchies—and the inequality they generate and protect—that many of us try to overcome did not exist before agriculture.

Again, Jackson's argument is not that every group of people in every location on the planet was bound to evolve in the same way once they adopted agriculture; variations in geography, climate, and environmental conditions produced different pathways. But those who captured the energy stored in surpluses of grain had a new way to accumulate wealth and power, and once those forces were loose in the world, people who made those choices would over time dominate those who didn't.

In the Mediterranean region, the ancient Greeks undermined their civilization's stability through deforestation and soil erosion. The civilizations of Mesopotamia struggled with soil erosion and salinization. The mighty Roman empire exhausted soils around the region, pushing into new areas to obtain the grain needed to feed people and then depleting those regions' soils. In ancient civilizations from central Mexico to China, the story of agriculture is also the story of erosion.

Agriculture leads to expansion of population, and more people means expansion of agriculture to feed them, leading to farming more marginal land, leading to more soil erosion and degradation. Technological advances can maintain, or even increase, yields in the short term, but the long-term problem is clear: There is no substitute for soil. Jackson has tried to impress upon people the importance of soil by pointing out that . . .

The Plow Has Destroyed More Options for Future Generations Than the Sword

Whatever one's preference in faith tradition, most people in the West know the key phrase from this verse in the book of Isaiah in the Hebrew Bible: "And He shall judge among the nations, and shall rebuke many people: and they shall beat their swords into plowshares, and their spears into pruning hooks: nation shall not lift up sword against nation, neither shall they learn war anymore" (Isaiah 2:4).

Jackson emphasizes that for all the destructiveness of war, the losses of soil and soil fertility are more devastating. Although the pain and suffering caused by wars—millions of people killed and wounded, societies shattered, cultures crushed—should not be minimized, the effects of a war are relatively short term. Populations can return to previous levels, societies can rebuild, cultures can be revived. Soil erosion poses a longer-term threat once we realize that "soil is as much a non-renewable resource as oil," as Jackson puts it.

Nature produces topsoil at the rate about 1 inch every 500 to 1,000 years. When wind and water sweep over a plowed field, soil erodes in an instant. Worldwide, about 23 billion tons a year erode beyond replacement levels, which is about 0.7 percent of the world's soil inventory.[1] Take a minute to do the math, and it's sobering to realize that at the current rate of erosion there would be no topsoil in about 150 years.

All ecological degradation destroys options for future generations. A hotter planet with fewer species and greater levels of chemical contamination—all of those crises impose burdens on people yet to be born. And all those stressors on the future will be magnified in a world with less, and less fertile, soil. Ironically, what the plow has destroyed increases the likelihood of war as people fight over limited resources.

Again, to be clear, Jackson does not contend that agriculture has been equally destructive in all places at all times, only that agriculture unleashes the potential for ecological degradation. That connection has been repeated across the globe regularly enough that it is surprising that most people aren't aware of it or don't seem to much care. Since annual grain crops (cereals, oilseeds, and legumes/pulses) take up about 70 percent of global acreage and produce roughly the same percentage of human calories, it is surprising how little public attention is paid to the crisis compared with the public conversation when war looms.

One obvious reason that politicians and the public do not attend to the crisis is the apparent success of contemporary agriculture. In the United States, there is an overwhelming abundance of food, and worldwide, there is also a surplus. All that food obscures . . .

The Failure of Success

Jackson does not use this phrase to describe the inequality in the distribution of food. About 800 million people today are undernourished, which is the result of poor people's lack of money to buy food, not the lack of food. That's an obvious failure, but Jackson uses "the failure of success" to point out that in our success at expanding our capacity to produce food, we have failed at a deeper level.

Yields began to increase dramatically in the twentieth century, especially in the post–World War II era with the widespread adoption of Green Revolution technology—petrochemical-based herbicides and pesticides and synthesized nitrogen fertilizers, along with more sophisticated machinery. Increasing yields to keep up with population growth seems to be a resounding success. Yet with today's success, Jackson reminds us, comes ecological destruction—today and tomorrow, on and on.

The fields of annual grains that require tillage every year are typically monocultures—one crop grown with little or no variation. These yield increases have led not only to increased soil erosion but also to the chemical contamination of land and groundwater, growing dead zones in oceans, and species extinction. The insecticides and pesticides necessary in these monoculture fields pose a threat to famers and farm workers. They contaminate groundwater that rural people drink as well as water downstream that cities draw on for drinking water.

To focus on just one of these problems, consider all the ammonia fertilizer used in the US agricultural heartland to boost yields of corn, soybeans, and wheat. Depending on the crop and conditions, only about half of that nitrogen is taken up by the plants, with most of the rest washing away with soil into streams and rivers, eventually into the Mississippi River and down to the Gulf of Mexico, creating a "hypoxic zone," the technical term for areas in the ocean that have such a low concentration of oxygen that marine life suffocates. The record for the size of the dead zone was 8,776 square miles in 2017, greater than the size of Massachusetts and up from the five-year average of 5,770 square miles.[2] Where we find modern agriculture around the world, we find dead zones.

Petrochemicals are designed to destroy plants that compete with crops for nutrients and water. As more and more land is converted to agricultural uses, species extinction increases. Rainforest is cleared to plant vast soybean fields in Brazil, and countless animal and insect species—some of them not yet charted by scientists—are eliminated.

And let's not forget climate change. The Millennium Ecosystem Assessment,[3] launched in 2001 and published in 2005, reported that agriculture directly contributed to 14 percent of global greenhouse gas emissions in 2005, which doesn't count the effect of deforestation to clear land for farming. The US Environmental Protection Agency put agriculture's share of greenhouse gas emissions at 9 percent in 2017.[4]

Jackson says that when he would go into too much detail for too long, his kids would say, "OK, Dad, got it." This bill of particulars to indict industrial agriculture for ecological crimes could go on forever, but with this short summary, we've got it. The short-term success—the result of a kind of "chemotherapy of the soil," as Jackson calls it—masks the long-term failure.

On second thought, maybe we've got it, but Jackson has more to say, both about the nature of the problem and the potential solutions.

3

Taking Nature Seriously

Many people have pointed out that it is a mistake to speak of human beings as separate from "nature," given that we are carbon-based like all other life forms and that no matter how much some people live in high-tech cocoons—which can feel distinctly unnatural—we are, and always will be, part of nature.

With that caution in mind, Jackson still finds it helpful to talk about the importance of nature, of being able to distinguish between the human-built and the nonhuman, between processes that have been ongoing on the planet for millions of years and processes we have introduced only recently. Here's one way that Jackson has captured this:

The best symbol for nature is a circle. Agriculture squared the circle. Industrial agriculture broke the square and stretched it into a line.

Nature is, indeed, circular—natural ecosystems feature material recycling, and in a healthy ecosystem there is no waste. What goes around literally comes around. Agriculture, no matter how carefully it has been practiced by responsible farmers, cannot replicate the elegance and efficiency of that natural circle. Industrial agriculture appears to be a line that is heading up on a graph—measured by ever greater yields— unless we consider the ecological costs. A full-cost accounting for industrial agriculture indicates that the line has been heading downward, toward an eventual reduction in productivity. If industrial agriculture is unsustainable—and all the evidence points to that conclusion—then we can't escape a simple fact: Unsustainable systems can't be sustained.

Perhaps the single most compelling example of the Alice-in-Wonderland nature of industrial agriculture is the concentrated animal feeding operation (CAFO), the technical term for meat production that concentrates hundreds or thousands of animals—chickens, turkeys, pigs, cows—in a factory setting.

In nature, plants and animals evolve to use resources in ways that create balance that can continue for extended periods; nothing is wasted. On a small farm, livestock are a source of meat and milk, and their manure provides fertility for the crops in the fields. Bad farming can degrade soils, but with diligence the human-managed landscape can maintain a rough balance in the squared circle.

In a CAFO, that manure collects in such quantities that not only is the potential fertility lost, but it becomes a toxic waste problem. The "success" in producing cheap meat masks the ecological and social failure, as both workers and neighbors suffer the effects, while animals are reduced to machine-like status and then treated with less care than machines. The elegance of the natural circle is gone. The struggle to live within the agricultural square is gone. The industrial system extracts the resources and exploits the animals until it is no longer possible to make money.

Still unsure about this critique? Not convinced of the failure embedded in our temporary success? Jackson argues that so many remain in denial because we are stealing, what the dominant culture considers to be . . .

An Acceptable Theft

An unsustainable system has been able to continue for so long, Jackson argues, because of a kind of theft that has been deemed—at least by those in power—an "acceptable theft."

Take our own backyards as an example. Jackson asks gardeners, when they speak of building soil, how much of the fertility that produces those plump, juicy tomatoes is taken from another place, coming from compost bags that were bought at the nursery? If gardeners create their own compost, how much of the organic matter that went into that compost actually came from that patch of land? That's not to run down gardeners, whose commitment to their soil is admirable, but rather to be honest about the sources of fertility.

The earliest thefts came when one group exhausted its own soils and made war to take the land of others. That kind of theft is not confined to ancient history; as recently as the late nineteenth and early twentieth centuries, colonial powers commandeered the agricultural productivity of the global South. The British conquest of India led to famines, as Indian wheat was shipped to Britain while Indians starved.[1] In the same period, when Europe was facing a fertility crisis, massive amounts of guano and sodium nitrate (also known as Chile saltpeter) were mined in South America and its offshore islands. Today, China's large-scale purchase of farmland in Africa will change the ability of that continent to feed its people.

The fertility for agriculture can be stolen not only from other places and other people but also from the future. Ever since the invention of the Haber-Bosch process in the early twentieth century, agriculture has depended on synthesized ammonia (NH_3) fertilizer created from atmospheric nitrogen and hydrogen, using natural gas as the primary feedstock. That industrial fertility is at the heart of the Green Revolution and yield increases, but it's merely another theft that those in power deem acceptable. In this case, the costs are borne not only by land and people abused in the extraction of the natural gas and other fossil fuels needed for the industrial process but also by people who will live with the effects of global warming and general ecological degradation in the future.

Imagine a neighbor who owns a big house, drives expensive cars, and boards a private jet to vacation. We might be tempted to envy the neighbor's apparent success, but if we found out the glamorous lifestyle was financed by theft, would we not reconsider our judgment? Jackson suggests that we do the same in regard to industrial agriculture. A truly sustainable agriculture that is capable of regenerating soil and fertility without theft will require a new standard, which Jackson captures in the phrase . . .

Nature as Measure

With more than 7.5 billion people on the planet, we cannot abandon agriculture without condemning billions of people to misery and starvation. If the standard for evaluating industrial agriculture—increased yields produced today with little thought about the future—is inadequate, what is the standard for ecological agriculture?

Jackson recalls one crucial moment in his graduate education in 1965 in his genetics lab at North Carolina State University. While Jackson was sitting behind his microscope one evening, looking at chromosomes, his major professor, Ben W. Smith, came into the room and offered this pronouncement: "We need wilderness as a standard against which to judge our agricultural practices." Smith left before Jackson could reply, but the thought stuck with him.

A decade later, after his return to Kansas, Jackson was reading an alarming General Accounting Office report on soil erosion at about the same time he was taking students at TLI out to study unplowed Kansas prairie. With Smith's words echoing in his mind, he pondered how the wheat fields of his home state could mimic the structure of the prairie, where grasses grew without soil erosion or chemical contamination. The obvious difference: Those grasses were perennials with deep roots growing in polycultures, not annuals with shallow roots growing in monocultures. In more technical writing, Jackson and TLI scientists have explained the challenges of breeding perennial grain crops, but most important here is the simple but revolutionary idea of pulling agriculture back toward nature's circle.

Jackson began to discover farmers and philosophers who had explored this idea of modeling human practices on natural systems and paying careful attention to the differences in landscapes, such as the eighteenth-century English poet Alexander Pope, who suggested that we "Consult the Genius of the Place in all."[2] In 1943, Sir Albert Howard published *An Agricultural Testament* and advocated "the Law of Return," the recycling of all organic waste material to build soil, observing that we should farm like the forest, for nature is "the supreme farmer."[3]

Jackson is not naive—there are social and scientific challenges to recalibrating agriculture. The 7.5 billion people on the planet all want to eat today and tomorrow and the next day, and we cannot move to a new agricultural system based only on the possibility of sufficient production in the future. Meanwhile, there are people who profit handsomely from the current agricultural system, and people rarely give up profit without resistance.

Jackson is not naive, but neither is he resigned to letting industrial agriculture spend down the ecological capital of the planet until the last check comes back owing to "insufficient funds." No matter how clever humans can be, human cleverness has yet to prove more reliable than natural systems, and our long-term prospects improve to the degree that we accept this. An agriculture that takes seriously nature as measure—a truly sustainable agriculture—will take varied forms depending on geography, climate, and environmental conditions.

When fossil fuels and high-tech machinery can do so much work on the farm, nature as measure can seem old-fashioned, and in some sense it is. Nature has been honing its genius for almost four billion years, and in this case, age does bring wisdom we can trust. Jackson warns that our faith in our own cleverness blinds us to the fact that . . .

High Energy Destroys Information,
Biological and Cultural

Jackson reminds us that if we look at any natural ecosystem, we should see information. In the uncountable life forms in the soil and on the surface—everything from the bacteria to the plants and animals—is an extraordinary amount of information in the DNA that controls the growth and development of those life forms. The natural ecosystem is information rich.

Now look at a monoculture cornfield—endless rows of that single plant, and with far fewer varieties of corn as seed companies have narrowed the genome of corn for maximal profit in the market. Increasing yields means destroying as many competing plants as possible with petrochemicals, applied with a high-tech tractor burning diesel fuel, fertilized with synthetic ammonia instead of through natural processes. Countless critters, insects, and other life forms can no longer find a home there. All the energy that made that field possible has destroyed biological information. Energy "solves" problems of yield, weeds, insects, and pathogens and along the way reduces biodiversity. The monoculture field is information poor.

Some of that biological information may return someday. Some of it is lost forever.

High energy can also destroy cultural information. A society that operates on contemporary sunlight has an incentive to be frugal and creative. When lots of fossil energy is available, practices that are virtuous in a low-energy world are typically abandoned for the "comforts" that dense energy provides.

Jackson uses the example of one of his grandmothers, who was known in the community for her expertise in scraping the intestine of a hog after butchering to get the high-quality lard without puncturing the intestine, which could then be used for making sausage. Those skills evaporate quickly when fossil energy brings to the market inexpensive commercial products to use in baking and sausage making.

Some of that cultural information may return someday. Some of it is lost forever.

For decades, Jackson has been asking, "What happens when the oil wells run dry?" We'll be scrambling to recover genetic variation in crops that will be important in a post-petrochemical agriculture, and we'll honor those with the cultural information about how to keep yields up in a low-energy world. The shift from industrial agriculture will require not only people who know things but also more people on the land, increasing what Jackson calls the ...

Eyes-to-Acres Ratio

Farming once was the dominant economic activity in the United States—about 90 percent of the labor force worked on farms when the country was founded. By 1900 that was down to 38 percent, and today only 2 percent of Americans are employed in agriculture. Meanwhile, the average size of a farm in the United States increased from 203 acres in 1850 to 461 acres in 1990.[4]

As the amount of fossil fuels and number of machines increase, the number of people needed on the land decreases. Huge tractors and combines burning diesel fuel can plant;

apply chemical fertilizers, herbicides, and pesticides; and harvest with one person at the wheel instead of dozens in the fields. When industrial agriculture substitutes fossil fuels and machinery for human attention, "efficiency" increases while the health of the land declines.

Jackson argues that there is an appropriate eyes-to-acres ratio for every piece of ground that humans occupy, depending on geography. One person on a cattle ranch in the Flint Hills of central Kansas can manage a sustainable grazing operation over hundreds of acres, but a grain farm of the same size would require far more people to walk the ground and care for the land.

Jackson emphasizes that a call for people to return to the countryside to farm is not mere nostalgia but a practical necessity. As the true costs of industrial agriculture become harder to ignore, society will have to find sustainable ways to farm, and those ways will require more people on the land. As other sectors of the economy change and jobs in both traditional industry and the digital economy evaporate, this need for labor on the land will be not a problem but a solution to the unsustainable nature of agriculture as well as the whole society.

Jackson also does not romanticize farm work. As a farm boy who spent a lot of time weeding row crops, he can testify to the strenuous nature of hand labor. But he also speaks of the pleasures of working the land in the company of people for whom you have affection, of the way rural communities can get hard work done with resolve and pride. Modern culture has long presented rural youth with the not-so-subtle suggestion that the countryside is for the crude and uncultured—think of the snide contempt in slurs such as "hick"—and the action is in the cities, where you find the future.

What if the future is in the countryside? What do we do with the industrial city?

4

Too Many, Too Much

While agroecologists work on sustainability in agriculture in the countryside, we can't forget that today's urban landscapes are unsustainable. That's the case not only of the world's most sprawling "megacities" (populations above ten million), where slums expand without basic waste disposal systems, transportation systems cannot keep up with sprawl, and clean water is often scarce. It's also true of a tidy, well-managed city with fully functioning infrastructure. As currently constructed, cities are an ecological disaster.

The unsustainable nature of these forms of social organization has to do with the real cost—not in financial but in ecological terms—of the buildings, roads, cars, buses, trains, underground pipes, water and waste-treatment systems, and everything else required for the operation of large human settlements in the high-energy epoch. The energy and resources required to construct cities and maintain urban life cannot be calculated accurately, but it is clear that those cities are also based on an "acceptable theft" that cannot continue indefinitely.

Because cheap fossil energy— cheap, at least, in short-term calculations—has led so many people to believe in the possibility of unlimited growth, analyses such as Jackson's are often brushed aside as hysterical, outdated, and irrelevant in the brave new world gone digital. But Jackson refuses to buy into the belief that technology will solve all our problems, and one strategy for demonstrating our folly is to focus on "embodied energy," trying to make a cradle-to-grave assessment of the total energy required to produce any good or service.

The embodied energy that automobile drivers consume when starting their cars, for example, is not just the fuel that they pump into the gas tank and burn on a trip. We also have to factor in the energy used to manufacture the car, from the mining of the metal to the final assembly. We have to include the energy that was consumed by the corporation in designing and marketing the car—some percentage of the energy used to build offices, fly executives to meetings, issue payroll checks, and perform every other activity related to that car. And then there's the embodied energy in the highways, traffic signs, and on and on.

When the ecological costs of all that activity are subsidized by the public through the government's underwriting of the fossil-fuel economy and/or pushed off onto future generations, then the money consumers spend to acquire and maintain a car is deceptive, and a calculation of the embodied energy tells us more. It helps us better understand what Jackson calls . . .

The Scaffolding underneath Industrial Society

If we routinely used "full-cost accounting," our understanding of the costs of our actions would change dramatically. We would train ourselves to see all the materials and energy that are required for our consumption. Take a simple example of eating an orange in Jackson's home state of Kansas. Oranges don't grow in Kansas, which means that when shoppers doing full-cost accounting reach for an orange in a Salina grocery store, they would think of the truck or rail car that brought that fruit north, the highways and rails on which the oranges traveled, the equipment in the orchard, the infrastructure of the grocery store, and so on. It is difficult to identify, let alone measure, all the energy required to make it possible to eat an orange in Salina, which is a fairly simple example.

Even more challenging would be quantifying the embodied energy of a cell phone. Trace every bit of energy in every component of a cell phone, at every stage of the process, from the tools used in mining the rare-earth metals to the human and ecological costs, present and future, in disposing of the phone when the new model arrives. The conclusion: We don't know the true cost of our living arrangements.

Jackson and colleagues at TLI tried to make such a calculation in the Sunshine Farm project. In the 1990s, on 150 acres on the TLI grounds, staff members grew organic vegetables and pastured cattle. The late Marty Bender went to great lengths to measure the embodied energy it took to produce food that relied as much as possible on contemporary sunlight, trying to account for all of the energy that went into everything, down to a single bolt in the biodiesel-powered tractor, from the mine to manufacture to transport. Jackson said that even with Bender's extraordinary dedication to the task and expertise in measurement and statistics, they learned quickly that an accurate assessment was impossible. The farther away from the farm they got, the harder it was to estimate, let alone calculate precisely, that embodied energy. That should give pause to anyone claiming that a particular practice with particular tools is "sustainable" because we cannot take into account all the "scaffolding" necessary to make the industrial world work.

Put another way: We should strive to make sure our boundaries of consideration match the boundaries of causation, even though we know it cannot be done. The boundary of consideration is drawn around what we choose to study, such as the agricultural production on the Sunshine Farm. The boundary of causation is everything that affects the object of study. The boundary of causation in studying the Sunshine Farm would have to be drawn far beyond the farm, reminding us that we lack the capacity to first, identify the proper boundary of causation and second, understand the myriad interactions within that boundary.

For years, Jackson has been using the scaffolding metaphor to nudge people to think about those boundaries, even though he knows it is not quite right. A scaffolding reminds us that the construction, maintenance, and repair of a building require a lot of materials, tools, and workers that we do not see when we walk in. The problem with the metaphor is that a scaffolding is removed when the building is finished—it is not a permanent part of the structure. But all that embodied energy Jackson asks us to consider is permanent; it is energy that has been used. We have to pay attention not only to the scaffolding but also to what is embedded in the foundation and frame of the building.

Until Jackson comes up with a better metaphor, he makes do with this one, hoping that it will train people's attention on what he calls . . .

The Extractive Economy

In some sense, every organism extracts resources from the ecosphere to live and thrive. Plants could be said to extract sunlight radiating to Earth in order to create foods out of carbon dioxide and water. Animals extract those plants out of the ground to live. We humans have long taken stone from landscapes to make tools. Everything that lives does so by taking things from the ecosphere.

But we moderns do a lot more extracting than any other organism, a process that began with agriculture, which

Jackson describes as a mining of the carbon of the soil, the first of those five carbon pools. The problem with our extraction, of course, is that some of the resources we extract are nonrenewable, and we are extracting renewable resources at rates beyond replacement levels. And all this extraction creates wastes that ecosystems cannot process. We are running out of lots of materials essential to the modern economy and creating a toxic environment in which to cope with the shortages.

Economists like to talk about substitutability, and the techno-optimists assume that when we run out of one resource, we will find another that can take its place. Running out of guano for fertilizer? Don't worry; Haber and Bosch eventually came up with a replacement. Toxic waste got you down? Don't worry; we'll invent a process to clean things up. That works until it doesn't work any longer, and it is folly to assume that the human desire to find ways to continue unsustainable extraction means that we can always find a way.

The task of facing reality is made more difficult by contemporary capitalism's fanatical commitment to the idea of unlimited growth. Capitalism is not the first system of unsustainable extraction, but it appears to be the most reckless and destructive. Short-term success in productivity—made possible by the extraordinary energy made available by the extraction of fossil fuels—has made it easy for people to believe that "more and more" can go on "forever and ever."

Jackson is one of a growing number of critics who know that the ecologically destructive nature of contemporary society means we cannot avoid a tough question: What is the size of a sustainable human population? Before venturing a guess, Jackson wants people to think about two different populations . . .

People and Their Things

If people ask about Jackson's opinion of the "population problem," he points out that the question makes no sense without asking about the level at which those people are consuming. It is not just how many people but how many things they have. Thinking of his longtime friend Leland Lorenzen, who lived in an eight-foot-by-sixteen-foot shack on about $800 a year, Jackson would say, "The world can support a lot of Lelands but not very many Donald Trumps" (that was long before Trump was elected president, back when he was merely the poster child for conspicuous consumption).

Talking about population control is difficult, in part because some people who advocate for population-control policies seem, either explicitly or implicitly, motivated by racism, ethnocentrism, and fear of global migration. Focusing not just on people but on their things, it becomes clear that the greatest threat to ecological sustainability is affluent people in the First World, not poor people in the Third World. Birth rates in the developing world matter, but Jackson's vision is broader: "We must practice contraception *and* contrasumption."

For Jackson, that doesn't mean simply searching for greater energy efficiency in the things we use, given the "Jevons paradox." In his 1865 book *The Coal Question*, English economist William Stanley Jevons demonstrated that increases in energy efficiency in production lead to more, not less, consumption, what is sometimes also called the "rebound effect." In other words, technological innovation without a collective commitment to reduced consumption doesn't get us out the hole but digs it deeper. Our carbon-seeking nature gets the best of us, and we have trouble recognizing when enough should be enough.

So improving efficiency is important, Jackson agrees, but more crucial is a sense of sufficiency—when less-and-fewer is sufficient to meet our needs. It's true that any two people might well disagree on what is a genuine need versus mere desire, but Jackson suggests that we should be able to agree that more-of-everything is a dead end.

To be clear, Jackson does not argue that the problems of the world can be solved by individuals deciding, on their own, to consume less. His focus is not on individual lifestyle choices but how we understand ourselves as people in communities—what does it mean to be human, and how do such humans live decently in community? When Jackson articulates his "mill-around" philosophy—"How do we mill around, amuse ourselves, and live cheaply until we die?"—he is imagining a world in which everyone lives cheaply in a community that doesn't allow living large.

Reducing the human population and curbing consumption is no small task, especially in an economic system that equates happiness with consumption. Proposals for what is called progressive change routinely invoke the public interest. But watch out, Jackson warns, for an enemy that lurks all around us . . .

"Public: Enemy #1"

In 1930 the *Chicago Daily Tribune* put Al Capone at the top of its list of "public enemies," and in 1934 J. Edgar Hoover declared John Dillinger to be America's first official "Public Enemy #1." Jackson has no affection for crime bosses or bank robbers, but he suggests that a much more dangerous enemy is the concept of the public, if used without careful consideration. It is time, Jackson says, to insert a colon in the phrase—"Public: Enemy #1."

Jackson is not demonizing members of the public—he has not developed a distaste for people. Instead, he is wary of the use of the "public interest" as a justification for policies that often turn out not to be in the interest of most of the people who make up the public. The "public interest" is an abstraction, and remember Jackson's skepticism about the demonic nature of "the abstraction without the particular."

Let's go back to Jackson's first cousin once removed, Dwight David Eisenhower (Jackson's mother was a first cousin to the general-turned-president, who was born in Texas but considered Abilene, Kansas, his hometown). Ike oversaw the construction of the interstate highway system, supposedly in the public interest. At the time, he sold it as important for national defense—remember that it was called the "National System of Interstate and Defense Highways" at a time when Cold War fears made anything associated with "defense" compelling. But beyond the hype, remember also that construction of the interstate highways required displacing communities, neighborhoods, and the people living in them. The abstraction of the public interest in moving cars and trucks faster overrode the particular interests of those people.

Did we-the-public really benefit from paving over so much of the country? We can get places faster—more efficiently, one might say—but are we better off because of that? Did making a car-based culture more efficient undermine support for more sustainable mass transit? How do we calculate the value of the locally owned shops of a small town that was once a stop on the older roads but disappeared when the interstate highways rerouted traffic? Are the chain restaurants on the interstate of the same value to people as a local diner? And are we better off because we can get between cities faster?

Jackson doesn't suggest that there are easy answers to these questions or that all people will agree on the answers. But when policies are pushed through in the public interest without concern for the interests of real people, it is easy for people in power to ignore those real people. People don't live in the public—we live in communities, and the health of those communities is too often ignored.

What makes it so hard to step back to critique all this unsustainable consumption? Jackson argues that it has a lot to do with our collective hubris, our belief that we're smarter than we really are.

The lectures during The Land Institute's annual Prairie Festival routinely fill all the seats in the "Big Barn," so many attendees bring lawn chairs to listen from the surrounding grounds.

Over the years, The Land Institute has acquired additional land outside Salina, Kansas, to expand the research plots that are crucial in developing perennial grain crops that can be grown in polycultures.

Kernza®, an intermediate wheatgrass, is one of the perennial grain crops being studied at The Land Institute and grown in research plots.

Wes Jackson addresses the 2018 Prairie Festival crowd in the "Big Barn," which he helped build with family, staff members, and interns in the early 1980s.

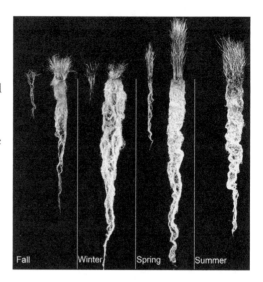

This comparison of annual wheat plants and perennial intermediate wheatgrass plants shows how deep the soil-saving roots of the perennials go into the earth.

Fall　Winter　Spring　Summer

Intern Maya Kathrineberg and research technician Marty Christians cut and bag heads of Kernza® for evaluation in The Land Institute's perennial grain breeding program.

Wes Jackson loads his vintage Ford tractor (a 1960 model, to the best of his recollection) onto a trailer for a trip to a repair shop. Jackson's brother, Elmer, brought him the tractor in 1974, when Jackson and his family were living on the property but had not yet opened The Land Institute.

An aerial view of The Land Institute in 1976, the year it was founded.

An aerial view of The Land Institute in 1989.

One highlight of Prairie Festival weekend is the tours of the experimental plots. Stan Cox explains his perennial sorghum breeding program during the 2017 festival.

Tim Crews, an ecologist and director of research at The Land Institute, takes measurements in a biculture field, where Kernza® was planted with alfalfa. When grown in a mixture, these plants can build an ecosystem that increases nutrients, soil organic matter, and biodiversity compared to a monoculture.

Interns at The Land Institute work in the fields, greenhouses, and labs with scientists, and they also have opportunities for informal chats with staff members. Wes Jackson shares a lunch break on the lawn outside the Research Building.

To mark Wes Jackson's retirement, members of the science staff at The Land Institute present him with one of the tools of the trade, a pollination apron, at the 2016 Prairie Festival.

Marty Christians, research technician for the Kernza® program, works with the intermediate wheatgrass plants in The Land Institute's cool-season greenhouse.

As the work of The Land Institute has expanded, so have the facilities. The two-decades-old original greenhouse, pictured here in 2017, is now used for perennial wheat and Kernza® research. A second greenhouse complex, for research on warm-season crops, legumes, and crop protection, began operating in 2018.

Aubrey Streit Krug (director of Ecosphere Studies), Wes Jackson, and Bill Vitek (director of the New Perennials Project) speak to scholars, artists, and activists during a break in the 2016 Ecosphere Studies conference.

The Land Institute's research plots just north of the main campus are nestled between the Smoky Hill River and a parcel of undisturbed native prairie called the Wauhob Prairie after Bessie and Loyd Wauhob, who in the early 1970s sold Wes Jackson the land that is now home to the institute.

5

Knowledge

Wes Jackson's trust in nature's wisdom over human cleverness is not a new idea—it goes back at least to Genesis and the story of Adam and Eve's banishment from the Garden, a story Jackson uses to talk about the possibilities, and limits, of human knowledge.

In chapters two and three of the first book of the Bible, God told the couple that they could eat freely of every tree in the Garden except the tree of knowledge of good and evil. Then came the temptation to reach beyond what humans know how to know, and after the couple ignored that warning, they were denied access to the tree of life. God sent them on their way to tough it out on their own, and ever since then things have been hard.

Jackson sees agriculture as the moment that humans started believing they could play God, intervening in ecosystems in ways that would eventually cause lasting damage. Leaving behind the misogynistic Eve-blaming interpretations that have done so much harm, the Genesis story is simply a warning against hubris—a reminder that we primates with the big brain should not overestimate the abilities that big brain bestows upon us. Human cleverness can lead us to believe that we are gods of a sort, that we can control the world since we got access to the tree of knowledge of good and evil, which came after agriculture unleashed what we call "civilization." While we fuss about our own brilliance, nature's wisdom and the tree of life endure, with natural forces that govern ecosystems sustaining the diversity of life on the planet without any help from humans.

For Jackson, the lesson of that story is not that people are better off stupid but only that we should resist the temptation to believe we are godlike and can manage our manipulation of the complexity of the world as well as nature does. The larger living world of which we are a part—the tree of life—was adequate to sustain us in our gathering-hunting economies, but we weren't satisfied. Now we are dealing with the consequences of our incompetence, our inability to fully understand and control our interventions into those ecosystems.

Can we reverse course, head back to the Garden, forgo the knowledge of good and evil, and rely on the tree of life once again? Well, there's the small problem that God "placed the cherubim, and a flaming sword which turned every way, to guard the way to the tree of life" (Genesis 3:24). But even without a flaming sword to repel us, there's no way to go back. As Jackson puts it, "Too much water has gone through the turbines to give up on the tree of knowledge." With soon-to-be eight billion people on the planet, returning to a preagricultural economy of gathering and hunting just isn't possible. There are too many people and too expansive a human infrastructure to imagine it's going to be an easy stroll back to a sustainable relationship to the ecosphere.

Jackson says we are going to have to do some negotiating with God (using that term metaphorically). We can approach the cherubim and ask to be allowed to keep our knowledge-seeking capacity but pledge that we will never again let it threaten to choke off the tree of life. We will agree to abandon the delusion of godlike omniscience for another chance at living with, rather than at the expense of, the larger living world.

Other traditions, including those of many indigenous societies in the Americas, offer similar wisdom about restraint and limits. Jackson makes it clear he uses stories that come from the traditions into which he was born in order to speak to the dominant culture. For Jackson, a good first step toward proving our sincerity would be to embrace an . . .

Ignorance-Based Worldview

Again, to be clear: Jackson's advocacy of an awareness of our ignorance is not an endorsement of being stupid but rather a recognition of the limits of human cognition. Given the complexity of the world, we are—and always will be—far more ignorant than we are knowledgeable, and Jackson suggests that "we go with our long suit."

The phrase "ignorance-based worldview" emerged out of a correspondence with Wendell Berry that led to the 2004 Matfield Green Ignorance Conference[1] and a subsequent book edited with Bill Vitek, *The Virtues of Ignorance*.[2] The basic principle is simple: Although human knowledge has expanded dramatically since the Enlightenment and the Scientific Revolution, especially during the high-energy industrial era, what we don't know still far outstrips what we do know.

Sensible scientists understand this, but even careful scientists can get drunk on knowledge and overestimate their brilliance. This isn't a problem just for scientists—this hubris lies in wait for us all, and because of these human weaknesses, we should be cultivating our capacity for intellectual humility. The appropriate course of action is caution: Do not assume that we humans have all the answers, understand how to ask the questions, or even know what the right questions are. We can intervene in the world in amazing ways—drilling down for oil and sending up satellites—but that does not mean we have the capacity to predict, let alone safely manage, the consequences of those interventions. Jackson makes this point by quoting Berry, who observed "how often and how recently our most advanced experts have been wrong about the future, and how often the future has shown up sooner than expected with bad news about our past."[3]

In short, human knowledge is not adequate to run the world. Jackson has suggested that it's as if we are stumbling around in a pitch-dark room that is full of priceless, and very breakable, antiques. In such a situation, lots of things get broken that cannot be replaced. Luckily, we have another choice—to recognize that we are fundamentally ignorant in the face of a world whose complexity we not only can't control but can't begin adequately to describe fully. As ecologist Frank Egler, a colleague of Rachel Carson, put it, "Ecosystems are not only more complex than we think, they are more complex than we can think."[4]

When we acknowledge the success of science in revealing physical processes heretofore unknown to us, the appropriate response should be not arrogance but the recognition of human limits. The more we know, the more we should be aware of what we do not know, and likely will never know.

This approach is not unique to Jackson[5] and is at the core of the "precautionary principle." If, for example, a corporation wants to introduce a new chemical into the environment, that principle places the burden on the company to prove it is safe rather than asking those who have concerns to prove it is harmful.[6] In an era of rampant science denial, this is tricky. To recognize what science can help us understand, we have to know the limits of science, but we also must be careful not to give ammunition to those who want to ignore science whenever its conclusions are inconvenient (such as deniers of climate change and evolution by natural selection).

Nothing in this view argues for abandoning all hope of knowing anything, for giving up on the search for truth, or for despairing that we can ever discover anything upon which we can act with confidence. It is neither a plea to renounce science nor a call to seek answers purely on nonrational grounds; instead, it is simply a call for intellectual humility. Jackson suggests that if we were to take seriously the idea that human knowledge is not adequate to run the world, then before embarking on a scientific or technological venture we would ask, "How many people will be involved? At what level of culture? Will we be able to back out?' Scientists, technologists, and policy-makers would be assiduous students of exits."

One of the greatest impediments to the needed cultural shift is what may be the most pernicious form of fundamentalism loose in the world today, which has taken hold throughout industrial society...

Technological Fundamentalism

There are many reasons to be worried about religious fundamentalism, but Jackson suggests that "technological fundamentalism"[7] is more dangerous.

In this sense, the term "fundamentalism" describes any intellectual, political, or theological position that asserts absolute certainty in the truth and/or righteousness of a belief system. Fundamentalism is an extreme form of hubris—overconfidence not only in one's own beliefs but in the ability of humans to understand complex questions definitively. Fundamentalism is not unique to religious people but is instead a feature of a certain approach to the world, rooted in mistaking limited knowledge for wisdom. Capitalists who believe that markets always know best (except when governments have to bail out companies that fail in the market) are economic fundamentalists. Patriots who believe the United States is always on the side of the angels (even when US leaders violate domestic and international law) are national fundamentalists.

Technological fundamentalists believe that the use of high-energy advanced technology is always a good thing and that the increasing use of ever more sophisticated high-energy advanced technology can solve our problems, including the problems caused by the unintended consequences of earlier technologies. These fundamentalists have no doubt that human knowledge is adequate to run the world. But to claim such abilities, we have to assume we can identify all the patterns in nature and learn to control all aspects of nature. That we so clearly cannot do those things does not disturb the technological fundamentalists' faith.

Contemporary ecological crises are evidence of our failure, but rather than step back and rethink our assumptions and reassess priorities, the dominant culture avoids the tough choices ahead by doubling down on the very project that brought us to the edge of the cliff. Finding themselves in a hole they have dug, technological fundamentalists argue for digging deeper and more furiously.

Jackson's interpretation of Genesis and caution about playing God seem particularly important in a world in which even some environmentalists are eager fundamentalists. "Nature no longer runs the Earth. We do. It is our choice what happens from here," says Mark Lynas in a book titled *The God Species*.[8] In 1968, Stewart Brand began the *Whole Earth Catalog* with the famous line "We are as gods and might as well get used to it."[9] Four decades later, Brand wrote that this suggestion had become an imperative: "We are as gods and HAVE to get good at it."[10]

An example: Technological fundamentalists typically are comfortable with pursuing nuclear energy, despite the potentially catastrophic consequences of a reactor accident and even though there remains no workable system of safely disposing of the waste. Folks endorsing an ignorance-based worldview would counsel against the development of an energy-production process that presents those risks and depends on the promise of some future solution to the waste problem. Technological fundamentalists assume the problems will be solved in time.

Need another example? How about chlorofluorocarbons (CFCs) and the ozone hole? CFCs have a variety of industrial, commercial, and household applications, including in air conditioning. They were thought to be miracle chemicals when introduced in the 1930s—nontoxic, nonflammable, and nonreactive with other chemical compounds. But in the 1980s, researchers began to understand that while CFCs are stable in the troposphere, when they move to the stratosphere and are broken down by strong ultraviolet light, they release chlorine atoms that deplete the ozone layer. This unintended effect deflated the exuberance a bit. Depletion of the ozone layer means that more UV radiation reaches the Earth's surface, and overexposure to UV radiation is a cause of skin cancer, cataracts, and immune suppression.

But technological fundamentalists might argue that our experience with CFCs is evidence for their optimism—humans figured out the problem and banned CFCs, and now the ozone hole is closing. These gases, which were once commonly used in air conditioning, were regulated in 1987 through the Montreal Protocol, which has reduced damage to the ozone layer. The oldest and most damaging CFC coolants have been largely eliminated from use, and the newer hydrochlorofluorocarbons (HCFCs) that took their place have little or no effect on the ozone layer. That's all true, but unfortunately we now know that the HCFC gases contribute to global warming. Scientists estimate that up to a quarter of all global warming will be attributable to those gases by 2050, so "the therapy to cure one global environmental disaster is now seeding another."[11]

So a reasonable question: If the dangerous HCFCs that replaced the dangerous CFCs are replaced by a new chemical that appears harmless, how long will it take before the dangerous effects of that replacement become visible? As

Jackson puts it, "How long until we discover the ozone hole equivalent?" There's no way to predict, but it seems reasonable to ask the question. Society didn't react to the news about CFCs or HCFCs by thinking about ways to step back from a developed world that has become dependent on air conditioning but instead continues to search for replacements to keep the air conditioning running.[12]

Rejecting technological fundamentalism and recognizing our intellectual limits does not mean freezing our understanding of the world at the current level but rather redirecting our attention as we realize that . . .

There Is More to Be Discovered Than Invented

Jackson—the farm boy who studied botany and genetics— says that science, buttressed by common sense, should lead us to one clear rule about our inventions: Don't introduce chemicals into our environments that our tissues have no evolutionary experience with. He counts it as a great failure of evolutionary biologists that contemporary society does not recognize this, to the detriment of all species, humans included.

When organisms evolve alongside a potentially toxic substance, tissues may be able to adapt, given enough time. But since World War II, tens of thousands of synthetic chemicals have been dumped into the world with little or no testing. We now know that some of those chemicals were clearly a mistake, and many more likely will prove disastrous. But we'll never know the effects—either alone or in combination—of most of these compounds. Even economists should be concerned about that since cost-benefit analysis is impossible when no one bothers to estimate the costs.

Jackson says that for too long we have been trying to figure out what we can get away with. In the pursuit of exploiting more energy to do more work for us, the modern economy tries to figure out how much damage we can do to ecosystems, plants, animals, and our own bodies and still slide by. Instead of always assuming we want more—whatever the good or service in question, in contemporary society the assumption typically is that more is always better—Jackson suggests that we start with a much different question: "What kind of world do we want to live in, and what will nature require of us to accomplish that?" Once we have begun to clarify our needs and recalibrate our desires, then we can go to the cherubim and say, honestly, that we have learned our lesson and are ready to live in right relation to the larger living world. We can promise that we will embrace our ignorance and leave behind technological fundamentalism.

Responsible knowledge-seeking, Jackson says, will shift our attention from invention to discovery. If we take nature as measure seriously, then more of our efforts will go to deepening our understanding of nature. Research on new technologies would not stop—at this point, we need better solar collectors and wind turbines. But instead of clamoring for the latest gadgets that widen the gap between human society and natural ecosystems, we would celebrate the discoveries that make possible low-energy living in stable, decent human communities.

Such a shift would require considerable change in every institution, including schools, colleges, and universities.

6

Schooling

Although TLI today is known primarily for NSA and research on perennial grain crops, Wes Jackson likes to remind people that it started as an alternative school. At the end of his second year of leave from California State University, Sacramento, Jackson and his family had to decide whether to return to the university or stay in Kansas.

The potential on the prairie won out over the steady paycheck, and Jackson and his wife at the time, Dana, launched a school in the fall of 1976 with eight students and modest financial help from a local supporter. With a house and some simple buildings on the Kansas property—which the family referred to simply as "the land," the origin of the institute's name—Wes and Dana sketched a program that split students' time between the classroom and hands-on work in pursuit of sustainable living. There were ups and downs along the way.

One down: The classroom building, with all the books and tools, burned to the ground just months into the project. The upside: When they offered the students a refund on their tuition, all refused and stayed to help rebuild. Jackson was out of the traditional educational system for good.

While Jackson admits that he was too restless to be a top student, he is grateful for what he learned from his time in a public school system, a small Methodist liberal arts college (KWU), and two large state universities (the University of Kansas and North Carolina State University). He taught in a Kansas high school and then at his undergraduate alma mater before heading to California, and all were positive experiences. Jackson speaks with great affection about his favorite professors and colleagues, and he respects and admires serious scholarship.

But he also has always bristled at the constraints that come with formal educational institutions. Trained in the sciences, Jackson is an avid reader of history and theology, and he collects friends who specialize in a dizzying range of subjects—from agrarian revolutionary movements in Russia and Eastern Europe to the philosophy of Alfred North Whitehead in addition to biology and ecology. Jackson likes book learning—calls to his friends, when sparked by a book, typically start "I want to read you something"—but he also has to regularly get out of his chair and into the wood shop, or on the tractor, or just take a walk to check on his cattle.

Jackson is, in a sense, intellectually and professionally homeless. "I really don't fit anywhere. I don't fit in genetics. I certainly don't fit in any university," he says, "and I don't think I would fit as a farmer." That lack of an easy fit might be behind Jackson's desire to reconfigure the way we organize knowledge and learning, which starts with ...

Asking Questions that Go beyond the Available Answers

Some of the most dramatic moments in intellectual history—such Galileo and his telescope, challenging the notion of an Earth-centered universe—come when the seemingly solid pillars of received truth start shaking. Isaac Newton challenged the mechanical philosophy of the day and gave us the law of gravity, a theory so radical at the time that even he had trouble believing it. Divine creation of species gave way to the insights of Charles Darwin (and, to be fair, Alfred Russel Wallace). Albert Einstein bent space and time in ways many of us still struggle to understand.

Most intellectual work does not produce such epoch-defining ideas, and most scholars work patiently to advance human understanding in much more modest ways. Jackson values the careful research that produces a slow accretion of knowledge—plant breeding, near and dear to his heart, takes a lot of time—but his concern is not so much with the answers as with the questions being asked. The road to guaranteed employment and status in the modern university is paved with lots of scholarly publications, no matter how dubious their value, which encourages researchers to focus on easily defined questions that can be answered by existing methods.

Jackson points out that if we keep asking only those questions that we know how to answer—or, at least, think we can answer—we are not likely to make much progress on the ecological threats that define our moment in history. Too often, fear that we cannot find an answer keeps us from asking the most important questions.

An example is the thinking that led to Natural Systems Agriculture. When first pondering why all our grain crops were soil-eroding annuals, the "experts" told Jackson that given how much energy perennial plants put into roots, they would never produce enough seed to make viable crops (remember that when we eat grains such as wheat and rice, we're eating the plants' seeds). No sensible plant breeder would waste time trying to either domesticate wild perennial plants or cross those perennials with domesticated annuals.

Luckily, Jackson has never been particularly sensible, and he and the TLI staff set to work on a project that was not guaranteed to succeed and, if success were possible, would take decades. There were plenty of failures early on, and the failures continue. But they have made steady progress, and now two plants—rice in China and Kernza® in the United States—are in limited commercial production.

Jackson is fond of saying, "If you are working on something you can accomplish in your lifetime, you're not thinking big enough." He's glad to have lived long enough to see some progress, but he's still cautious about proclaiming NSA a success. Researchers still face questions that go beyond the available answers—initial success in an ambitious project tends to give rise to new challenges—prompting TLI to widen the scope of inquiry, hiring people with expertise in soils, insects, pathogens, and other specialties in agronomy, ecology, and biology. Jackson is after what he calls a "yeastiness of thought" that comes when a diversity of folks come together to ponder questions that have no quick answer, or maybe no answer at all.

For Jackson, that means not only bringing scientists from different fields together but getting the humanities and social sciences into the conversation. Jackson would love to see a loosening of the disciplinary boundaries that structure the modern system of scholarship, which could help . . .

Drive Knowledge out of Its Categories

The post–World War II expansion of the US university system, in tandem with the civil rights and feminist movements, had a generally democratizing effect, allowing greater access for people who had once been locked out of higher education. That rapid growth in funding for teaching and research also led to increased specialization, some of which was necessary as knowledge proliferated. For example, since Jackson earned his PhD in 1967, entire subfields in genetics have emerged, developing highly specialized concepts and terms necessary to work in the field. The advances have been impressive.

But Jackson believes that the hyperspecialization of contemporary academic institutions has come at a cost. We know more, but are we striving to know the things that matter most? We know more, but are we putting what we know to good use? We know more, but are we wiser? Jackson's concerns are not anti-intellectual but prointellectual. He is simply suggesting that we think more about what the goal of intellectual life should be. Many have decried the focus on the bottom line in US universities—the research that matters the most is increasingly that which brings in the most revenue—but Jackson is just as concerned about the rarity of deep cross-disciplinary conversations, the lack of real engagement among colleagues.

What kinds of insights are lost when researchers' vision becomes too narrow? How can we connect the dots if all the different dot makers from different dot-making departments rarely talk to each other? Jackson believes that knowledge is produced in unscripted conversation as much as in books and scholarly papers.

The lineup of speakers over the years for the annual Prairie Festival reflects Jackson's preference for hybrid vigor. For example, the 2013 speakers included a philosopher/tree farmer who taught in a school of environment, an economist interested in what we can learn about our economy from ants, a scholar of religion who founded a sustainable development program, a plant breeder who wrote a book about what's wrong with air conditioning, the manager of an investment fund, and one of the country's most well-known food journalists. Jackson's talk that year explored how the David and Goliath story might be instructive for people working for sustainability (quite a lot, he argued, given that the odds didn't seem to be in David's favor, either).

For Jackson, tossing all these folks together makes for a fun weekend, but the larger goal is figuring out the shape of a sustainable future and educating young people to negotiate that landscape. If Jackson were in charge of the country's higher education, he would add one new program to the catalog . . .

A Major in Homecoming

Despite the proliferation of new departments and programs in universities, Jackson is fond of saying that there really is only one major: "upward mobility," training people to advance in the modern economy with no necessary connection to any particular place. That's true not only of young people studying business and finance. No matter what the subject, the assumption is that students will leave the university and chase the best opportunities, typically defined as those that pay the most or convey some other type of status. If you want to make something of yourself, the action is in cities, with the start-ups, using and developing cutting-edge technology. The advice is, go cosmopolitan.

Jackson offers young people another option. If you want to contribute to a decent human future, go home. By that, he doesn't mean everyone should return to where they were born and grew up, nor does he think everyone has to head for the countryside. Instead, he advises people to find a place where they can make a home and dig in for the long haul.

Community, Jackson believes, is where the action is. Hammering out public policy at the state, national, and international level is important, and Jackson does not advocate abandoning that work. But our evolutionary history makes us most at home in small communities, what Jackson calls civilization's version of the band-level societies we evolved in. Humans are constantly inventing new ways to live, of course, but life at that smaller level of social organization is what comes "naturally"—it helps reduce the species-out-of-context-edness of our existence. That is where Jackson believes we will have to rebuild in the low-energy world that awaits us.

Jackson wants a major in homecoming that could help students focus on the skills and wisdom people will need to make communities work, especially in the face of a capitalist economy that worships bigness and has helped to destroy communities, both rural and urban. Some of the instruction that will be needed is in how to get things done with less energy—food production that doesn't rely on cheap fossil energy; the domestic arts of homemaking that have atrophied in a consumer culture; tinkering to extend the useful life of industrial products when the throwaway world is over.

But just as important to Jackson as sustainable agronomy and engineering are the arts and humanities, the cultivation of a deeper sense of what it means to be human than we get from consumer culture. Jackson knows that the meanness that can surface so easily among strangers in a city doesn't magically disappear in smaller communities. How will we cope with scarcity? How can we fashion shared values through democratic engagement rather than authoritarian social control? How will we entertain ourselves without relying on an entertainment industry?

Jackson offers *The Odyssey* as an example of the books that students in a homecoming major would read. Homer's epic follows Odysseus on his quest to get home. That included seven years as a prisoner of Calypso on an island where, if he had only agreed to become her husband, he could have lived forever with endless pleasure. An attractive offer, but still Odysseus yearned for home, for the place he knew he belonged. *The Odyssey* is a reminder that the temptations of upward mobility have less hold on us when we know our place in the world.

This approach to education already exists, albeit on the margins. Students at "work colleges," such as Deep Springs College in California and Paul Quinn College in Dallas, study in traditional classrooms but also work to maintain the college buildings and grounds. The Wendell Berry Farming Program in Kentucky teaches students how to practice an ecological agriculture that also can be economically viable. And within more traditional universities, small experiments in crossing disciplinary boundaries and getting students into the community are popping up more frequently.

Jackson is neither romantic nor naive; he doesn't believe that a homecoming will take us back to some mythical golden age, nor does he think that a return to community will magically allow a kinder and gentler high-energy lifestyle. No matter where we choose to live, the bills of the agricultural and industrial revolutions are coming due, and that means major changes. It means humans face a unique challenge in the history of life on the planet.

7

"We" Have to Accept Limits

Humans have made a mess of things—by now that should be readily evident to anyone willing to face the avalanche of studies and statistics that describe the contemporary ecological crises—and still as a species we have not committed to a serious project to clean up the mess. That profound failure might lead some to misanthropy; it is tempting to conclude that the species is hopeless and just give up on people (though that would imply giving up on oneself, if one is part of the group "people," and it's not clear what that would mean).

Jackson suggests that instead we should go easy on ourselves, which may seem surprising given his critique of so many of our species' failures for the past ten thousand years. Should we not be tougher on ourselves, pushing each other to change not only social policy but also personal behavior?

Yes, we need to hold each other accountable. Yes, we have to hammer out rules of behavior to promote sustainability and justice. But let's be clear about just what we are asking of ourselves, how revolutionary the task before us truly is.

Remember Jackson's description of life: the scramble for energy-rich carbon. We have made a mess of things by becoming exceptionally good at what all organisms do—going after that carbon. We have gotten so good at carbon-seeking that if we don't change our ways, it's not clear how much longer the ecosphere will put up with us. *We* need to be both harsh on each other and kind to ourselves because *we* face a unique challenge—*we* must become the first species to impose limits on itself, the first species to curb its scramble for carbon by a collective act of will.

One question to ask before tackling that challenge is . . .

Who Is "We"?

Humans today do not all consume the same amount of energy, nor have all humans throughout history. When it comes to ecological destruction, "we" know that "we" are not all part of the same "we." There are important differences—among individuals within a single society and among societies—in energy consumption, resource exploitation, waste production, and overall contribution to ecosystem instability. Many of those differences are the product of violence, coercion, and fraud (whether legal or illegal), as some people take more with no moral justification. That must always be resisted.

But what if we were to create a more just world and eliminate such glaring inequalities? It is tempting to believe that from there the solution lies in simply identifying low-energy societies from the past or communities in the contemporary world that live a lower-energy existence and then building the political will to reproduce them in the dominant culture. But such an approach would be a dead end—not because it would be hard to build that political will (which is true enough) but because it misunderstands the nature of the problem. We can learn from the low-energy societies and experiments within conventional societies, but those good examples don't offer a strategy to move forward from where we are (high-energy, unsustainable) to where we need to be (low-energy, sustainable). First, the lives of billions of people around the world are dependent on high-energy infrastructure, and the necessary transition will, to say the least, not be easy. Second, we can't pretend that people will overnight find it easy to avoid the temptations of dense energy, individually or collectively.

True, the profoundly unsustainable nature of human economic activity today is to a great degree the result of the European conquest of the world over five centuries and a rapacious transnational corporate capitalism. But as Jackson reminds us, those temptations of dense energy are powerful for humans because we are organic creatures—human nature is carbon nature. While much of today's irrational consumption is driven by capitalist propaganda (that is, advertising and marketing), fossil fuels and other sources of energy also make people's lives easier in many ways that are not frivolous. There is of course variation in people's assessment of their needs, contrasted with wants or desires, but capturing and using dense energy for comfort and pleasure is not an activity unique to imperialists and capitalists.

To be clear: For those of us with a disproportionate share of the world's wealth who are responsible for a disproportionate share of ecological destruction, the argument for change is strongest. But in trying collectively to find a way out of the mess we've made, the assigning of different levels of responsibility for the mess is only a first step.

Renouncing First World dominance is a start, as is imagining a world beyond capitalism's obsession with growth and consumption. The end of those systems is a necessary but not sufficient condition for change. No culture has a plan for transitioning from an unsustainable high-energy global society of eight or nine or ten billion people to a sustainable low-energy society with a dramatically lower population. This is a new challenge with no road map available. Lessons from low-energy societies will undoubtedly be valuable, but there is no way to flip a switch and return to a previous era's living arrangements. No one person, ideology, or culture is going to provide us with a template for dealing with what lies ahead.

If we start with an awareness of the scope of the change needed and the lack of a plan, we can at least be clear about the direction in which we need to move, which takes us to the challenge of being ...

The First Species That Will Have to Impose Limits on Itself

For the past couple of decades, Jackson has believed that no meaningful progress is likely without a global cap on our use of carbon, along with a rationing system to ensure a just distribution of the resources we use. We all collectively will have to choose to live with less of just about everything—less energy burned, fewer metals mined, a dramatic decrease in gadgets—and enforce those limits through public policy. That's not exactly a winning campaign slogan for politicians, who typically run on a promise of "more." But Jackson isn't deterred. "A cap on carbon with rationing," he repeats, as often as possible. Why push so hard for what seems so implausible, at least in his lifetime?

If we want to plan rationally for our future, we have to assess the likely trajectory of not only climate change but other unfolding ecological crises. If we abandon the technological fundamentalists' faith-based claim that we will invent our way out of the crises, then we have to think about how to power down. We have to understand that our ten-thousand-year run of "success" in the scramble for energy-rich carbon—starting with our exploitation of the carbon in the soil through agriculture—is actually our greatest failure.

Jackson highlights the painful irony of being human in this moment: Our unprecedented success at getting at that carbon has allowed us to dominate the planet in a way unlike that of any other species. As a result, we must do what no other species has ever had to do, or could even contemplate doing: We must limit our carbon-seeking through collective willpower, recognizing that there's no guarantee it's possible.

When other organisms expand beyond sustainable levels, they are kept in check by natural processes involving predators, disease, and limited food supplies. Humans have temporarily transcended those limits through increasingly sophisticated exploitation of the five carbon pools, but this condition is temporary—unchecked human growth will be checked eventually, whether we do it ourselves or leave it to nature. If we want a decent human future, we must not wait for nature but impose limits on ourselves and each other. A cap on carbon gives us a shot at sustainability, while rationing makes social justice possible.

How will we do it? What is the appropriate level of the cap? How much energy can we afford to burn over what time period? What kind of rationing system will work best? And how can we implement any of this in a world full of people who still want more?

Don't ask Jackson for too many details. No one else has definitive answers, though his TLI colleague Stan Cox has done important work to lay out the scope of such a project.[1] But Jackson is a firm believer in making a commitment to achieving an outcome before we know how to do it. The bigger the problem, the less we know about how to solve it and the more important it is to make the commitment, even if there seems to be no solution within reach.

Is it rational to hope for a policy that no one can yet imagine? The better question: Is it rational to continue to avoid planning for the most likely future only because it is difficult?

It's tempting to blame this failure on, in no particular order, rich people, really rich people, people who desperately want to be rich, Walmart, Amazon.com, ExxonMobil, climate-change deniers, conservative politicians, all politicians, campaign-finance laws, imperialism, white supremacy, or capitalism. There's no shortage of targets, and all share in responsibility. But let's not forget the role of our shared human-carbon nature, our carbon-seeking nature. Some would say that if that really is our nature, we are doomed to fail. Jackson flips the argument: An honest assessment of these realities at least improves the odds for rational planning.

If rational planning at the national or international levels is impossible at the moment, remember Jackson's focus on communities. What can be accomplished at lower levels does not solve the problem but demonstrates that we are competent to run our own lives. In the absence of the political will to pass a cap on carbon with rationing in Washington or the United Nations, we can look for opportunities at home. Let a hundred flowers bloom, and be careful not to dismiss any idea too quickly. Jackson doesn't talk about how to "save the world"—we're the ones that need saving, not the world, he says—but rather how to limit the damage we do and try to make possible a "soft landing" for humans in the future.

At this point, we're all making it up as we go along. We can learn from people who have lived in low-energy societies, but those living arrangements do not offer a blueprint for how to power down, how to reduce the population of people and our things. We are facing a transition that has not been made before, and it's going to be a wild ride. But we stand a better chance of making it to the other side if we abandon the industrial worldview that got us into this mess.

8

A Creaturely Worldview

One of Wes Jackson's favorite phrases is "bear down," used when he comes upon an idea that he believes merits close attention. "We need to bear down on that," Jackson will say.

As Jackson gets older, he seems to speed up intellectually rather than slow down, a reflection of how much there is on which we need to bear down. The list includes ecological problems as old as soil erosion and as new as the Great Pacific garbage patch, all intensified by the realities of rapid climate disruption. Indeed, there is a whole lot of bearing down that needs to be done.

To increase our chance of coping effectively with these crises—the likelihood that we will bear down on the right questions in productive fashion—Jackson is careful to focus not only on specific problems but also on the relationships among the problems. That is why he regularly circles back to the need for a new worldview, a challenge to the industrial mind of the modern age. Jackson has argued that the break with nature came when people started farming, but the human assault on ecosystems—now routinely referred to as an existential threat—greatly intensified with the Industrial Revolution. The first step toward a sustainable future is to diagnose the industrial failure. If we are stuck with agriculture—7.5 billion people are not going to return successfully to gathering and hunting—we can make it less industrial and more creaturely.

My conversations and correspondence with Jackson about these matters led to a collaboration on the essay "Let's Get 'Creaturely,'" which was published on the website of the Post-Carbon Institute.[1] This essay, revised slightly from the version that appeared online, brings together many of the ideas laid out in this book.

Let's Get "Creaturely": A New Worldview
Can Help Us Face Ecological Crises

by Wes Jackson and Robert Jensen

No farmer has ever gone out to the barn to start the day and discovered that a baby tractor had been born overnight. For farmers who work with horses, the birth of a foal would not be surprising.

That observation may seem silly, but it highlights an important contrast: Machines cannot reproduce or maintain themselves. Creatures can.

The tractor comes out of the industrial mind, while the horse is creaturely. The tractor is the product of an energy-intensive human-designed system, while the horse is the product of an information-intensive biological process that emerges from earth and sun.

The implications of this difference are rarely acknowledged in the dominant culture, but we believe they are crucial to explore, especially with new political space opened up by the Green New Deal for discussing ecological sustainability and economic justice.[2]

In the short term, humanity needs to devise policies that respond in meaningful ways to today's multiple, cascading ecological crises (including, but not limited to, rapid climate disruption), which present risks now greatly accelerated and intensified well beyond previous predictions. If that seems alarmist, we recommend "World Scientists' Warning to Humanity: A Second Notice." More than 15,000 scientists from 184 countries warn: "We are jeopardizing our future by not reining in our intense but geographically and demographically uneven material consumption and by not perceiving continued rapid population growth as a primary driver behind many ecological and even societal threats."[3]

To put uncomfortable realities bluntly: In ecological terms, things are bad, getting worse faster than anticipated, leaving humanity with increasingly limited options. Everyone agrees that there are no quick and easy fixes, but we want to push further: Do not expect any truly sustainable fixes that will endure over the long term to emerge from the industrial mind.

That's why we believe it's crucial to discuss not only policy but the need for a new worldview, one that can expand our imaginations. The distressing realities of our moment in history need not be the end of our story, if humanity can transcend the industrial and get creaturely. As we deal with the 10,000-year-old "problem *of* agriculture" we're more likely to make meaningful progress by replacing an Industrial Worldview with a Creaturely Worldview.

Creatures—humans, other animals, plants, and microbes—are all products of a rich, integrated evolutionary history. Unfortunately, something in our big brains has too often led too many of us to see ourselves as set apart from the rest of the larger living world—to think of "human" as so different from "nature" that we believe ourselves to be separable from the ecosystems on which our lives depend. The Industrial Worldview, deeply rooted in this delusion, defines much of our day-to-day existence and suffocates our imaginations.

What if we embraced a Creaturely Worldview as a corrective? This would challenge not only the dominant culture but also some in the environmental movement who are committed to industrial thinking and its accompanying technological fundamentalism. The current debates about the Green New Deal might be more productive if everyone started by considering this question: Which provides a better standard for our choices, the Creaturely World or the Industrial World?

Before proceeding, an acknowledgement: We are not the first to ask this question, of course. In some sense, indigenous and traditional people who resisted the Industrial World have long advocated for a Creaturely Worldview. The Amish rejection of some of the products of the Industrial World reflects a faith in the creaturely. The agrarian writer Wendell Berry, one of our touchstones in this enterprise, has spoken of the costs to people and land in the countryside with the "change from a creaturely life to a mechanical life"[4] that accelerated after World War II. With that acknowledgement of our roots, back to the argument.

First, remember that the Creaturely World had a considerable head start. Creatures have been here some three-and-a-half billion years. The Industrial World has existed for only 250 years, about fourteen million times shorter. By linear comparison, that's roughly the difference between an inch and 220 miles.

We argue for the Creaturely based not just on time but more importantly on the greater creativity and efficiency of nature's ecosystems, compared with the limited vision and mixed record of human cleverness. The Creaturely World features self-organizing renewability (remember the horse and foal) emerging from the integrated structure of ecosystems—what we might call the "natural integrities." A tall-grass prairie ecosystem, for example, is not a random collection of species but the result of natural selection that produces species interacting with each other and with the abiotic world in ways that efficiently utilize the available resources. Glaciers, plant and animal migrations, fire, grazing, and climate all played a role in shaping that prairie, not to build a "perfect" ecosystem as if there were an actual designer at work, but creating a relatively stable home for life over long periods.

The Industrial World erodes those integrities, requires human attention to maintain, and is non-renewable. For the Industrial World to work, dismemberment of integrated nature is required.

The Creaturely World is *information-rich*; the genetic code of organisms stores enormous amounts of information. People routinely speak of living today in an information age made possible by digital technologies, but this human-generated breed of information is only a tiny fraction of what is found in the DNA of the Creaturely World. The fact that human inventions are relatively information-poor is typically obscured by our use of highly dense energy to compensate.

A perfect example is anhydrous ammonia as a source of nitrogen fertilizer for modern agriculture, the product of what energy scholar Vaclav Smil has called the most important invention of the twentieth century, the Haber-Bosch process.[5] Natural gas is the feedstock most often used to turn tight-bonded atmospheric nitrogen into ammonia. This industrial process "solved" the problem of soil nitrogen fertility and declining supplies of natural fertilizers such as guano. Unfortunately, after being spread over millions of acres of grain-producing fields, the surplus industrial nitrogen finds its way down the slopes and into the waterways until it meets the ocean waters, where it creates huge dead zones. On the way downriver, cities spend millions of dollars to get it out of drinking water, in some places failing so dramatically that people have to drink bottled water.

Haber-Bosch does its assigned job of increasing crop yields, but with a climate-changing cost: It uses fossil energy to generate the 200 to 400 atmospheres of pressure and temperatures of 750 to 1,200 degrees Fahrenheit required to produce ammonia from the natural gas feedstock. In contrast, the biological process of nitrogen fixation in various plants operates at four-fifths of one atmosphere of pressure and at ambient temperature, relying on twenty-one enzymes that are the product of the DNA code—drawing on the natural integrities of the Creaturely World. Thus the Industrial World's nitrogen production substitutes fossil-energy for information, disrupting ecosystems' integrities with a non-renewable process that contributes to ecological degradation, from the mining of the fuel to the acceleration of global warming.

This is an example of a larger rule: Ecosystems are far more creative than human systems. Consider a modern city, the product of the human-generated information used to build the housing, businesses, infrastructure, and transportation networks that allow millions to live in close quarters, often with exciting results (both constructive and destructive). All that excitement leads us to ignore the fact that these cities of the industrial age are made possible only through massive expenditures of fossil energy and other resources, some of which come from the other side of the planet. Meanwhile, natural ecosystems are home to a much more expansive variety of creatures living in far more complex relationships, requiring none of that fossil energy to maintain. Natural ecosystems can maintain themselves for countless millennia using only solar flows, while cities draw down millions of years of concentrated energy in a relative blink of an eye. Which model provides a standard for our future?

Here is an idea that is counterintuitive in the modern world: *Highly dense energy limits the human imagination.* Yes, all that fossil energy has subsidized a tremendous amount of science and art, expanding dramatically what we know about the world and building an expansive trove of stories about it. But rather than imagining how we might use that energy to build a sustainable future, we have rushed to use it in ways that enriched some quickly, impoverished others slowly, and left us facing a future that is speculative, not guaranteed. As we come to the end of the fossil-fuel epoch, as a species we seem to lack the collective imagination to break free.

Another challenge to the conventional wisdom: The Industrial World acts as if public policy is made by humans and those policies determine how we use energy, but it might be more accurate to say that *highly dense carbon, once unleashed, sets policy and drags us along.* If we set aside our belief that we humans are always in charge, we can ask a troubling question: Did we build the contemporary world by making choices about how to use energy, or did highly dense energy dictate the shape of the contemporary world, in which we make choices that have been constrained by the industrial mind? The choices we do make within the Industrial Worldview matter very much—we can opt for more or less destructive paths—but in the long run, it is the worldview that has to change.

Let's pause to answer a reasonable concern: Are the two of us zealots? Do we want to give up on everything humans have ever built? Are we calling for a mystical return to the Paleolithic tomorrow? No to all those questions. Are we proposing to "let nature take its course," and stand by while billions of people die in such a transition? Certainly not. Advocating for a shift in worldview is a plea for new ways of thinking, not a celebration of misanthropy. Rather than throwing up our hands in despair because imaginations have been so limited in the industrial era, we suggest that the dominant culture start identifying and attempting to follow the patterns of the Creaturely World—not an atavistic return to any particular moment in the past but rather attention to the lessons of evolutionary history.

An example is The Land Institute (TLI) in Salina, Kansas, where both of us have worked since 2015 with the Ecosphere Studies program. The term "creaturely" doesn't appear in the organization's mission statement, but the Creaturely Worldview informs its work.

For more than 40 years, TLI researchers and teachers have advocated for nature as the standard for grain farming, as they work to develop an information-rich agriculture that mimics the vegetative structure of an information-rich native prairie. This Natural Systems Agriculture project is developing perennial grain polycultures (grain crops that need not be planted every year, grown in diverse mixtures), a more creaturely approach to agriculture than the annual monocultures in industrial fields. In addition to reducing soil erosion, those perennial grains would sequester more carbon, and adding legumes to the mix sponsors biological nitrogen fixation, removing the need for the Haber-Bosch process and its accompanying emissions.

(An aside: There also are approaches to farming annual grains and other kinds of agriculture—often bundled under the term "agroecology"—that are more creaturely than the current industrial agriculture. We support these efforts, alongside the work to develop perennial polycultures, as part of a sustainable agriculture movement.)

In this work, TLI staffers recognize that every day they use the products of human cleverness and industrial society—booting up computers, carrying tools to the research fields in pickup trucks, transferring pollen in three natural gas-heated greenhouses, and burning fossil fuel to warm labs and offices—all in the hopes of developing crops that can endure without all the trappings of the Industrial World and make possible a transition to truly sustainable agriculture.

The big test that's coming: Once we have these new species and varieties, will growing them at the scale necessary to feed people require maintaining the industrial infrastructure that brought them into existence? We believe the answer is "no," that their creatureliness will persist without a need for human intervention. These species could be grown by people who never touched a computer, and could be maintained without the artifacts of the Bronze and Iron Ages. The Industrial World can't say that of many, if any, of its achievements.

We realize that we cannot get to perennial polyculture agriculture without some of the tools of the Industrial World. We continue to use fossil fuels, though over time we hope we can supply more of the power for this transition period with solar and wind technology. But we cannot be naïve about "renewable" energy, and the Creaturely Worldview can help us understand why.

First, the easy part: No combination of renewable energy sources[6] can power the existing Industrial World. Rational planning must include not only replacing fossil fuels with renewables but also dramatically reducing our consumption. If we are using renewable energy to try to produce enough electric cars to continue our current transportation system, for example, we are only digging the hole deeper, not finding ways out. An enforceable cap on carbon at the mines, the wellheads, ports of entry, and forests seems necessary, which means we'll also need a fair rationing system.[7]

Second, the hard part: There are limits to renewable energy technologies' ability to replicate themselves. At the risk of unnecessary repetition: The Industrial World is not self-renewing. Working against instead of with the efficiencies inherent in natural integrities means that a considerable amount of energy that so-called renewable technologies produce must go into mining and manufacturing the non-renewable materials required for that infrastructure. That's a losing game. Wind turbines and solar collectors built with fossil-fuel infrastructure will not be easy to maintain or reproduce when that fossil energy is gone.

What does that Creaturely Worldview have to offer here? We can begin by scrutinizing proposals under the Green New Deal umbrella, most of which embrace "green-energy cornucopia"[8] thinking that keeps us entranced by the industrial mind's illusion that we can sustain unsustainable living arrangements. As even many of its supporters understand, the problem is not that the Green New Deal is too ambitious but that it is not ambitious enough. Virtually all politicians, and even many who identify as environmental activists, embrace a growth economy and techno-optimism. As difficult as it is in mainstream political circles, we must challenge those dogmas and imagine a transition to a more creaturely economy.

A Creaturely Worldview requires dramatic changes in social and political arrangements. One obvious shift would be reducing the size of farms and increasing the farm population, recognizing that we can better feed ourselves by relying on a sufficiency of people rather than on capital and dense energy. Repopulating the countryside would require something like a new Homestead Act, creating an opportunity to correct the extermination and exploitation of both peoples and ecosystems that was woven into the first version in 1862.

Of course there's a clear need for short-term industrial productivity as the transition unfolds, and there may be a place for the Industrial World in our future—but only if it is clearly subordinated to the Creaturely World. Wind and solar energy are a good example of that: We'll need them in the transition period, but our reliance on them should shrink (unless, by magic, wind turbines and solar collectors start having babies) as we get closer to the creaturely goals.

Obviously, a Creaturely Worldview doesn't have all the answers to all problems. A worldview doesn't solve problems but rather shapes the way we understand questions and guides our search for answers. While articulating a vision for the future we draw upon our imaginations, which cannot be divorced from our evolutionary history. The way we describe the future is always partly new and partly rooted in that history. With that in mind, we might think of the Creaturely World as a kind of New Paleolithic, the next step forward after the ephemeral fossil-fuel epoch has run its course.

In such a Creaturely World, there will be less of many material things that many of us (including the authors of this article) have grown accustomed to, but potentially more of the one thing the Industrial World could never produce on its own: a sense of being at home and cherishing our origins in a universe that is not just a place but also a story.

We are but one part of that story, and our place in it should feature earth as our creator, our defender, and—with proper restoration of the Creaturely—our redeemer.

Conclusion:
Spontaneous Elaboration

One of Wes Jackson's favorite quips about his intellectual work comes from a conversation with his brother, Elmer. Noting that Jackson is always quoting others in his writing, Elmer asked, "Don't you have a mind of your own?" Jackson happily conceded that he did not.

"I don't know what I think until I talk to my friends," Jackson likes to say. He has plenty of ideas of his own, of course, but he is serious about how much he relies on friends and colleagues, especially those with expertise beyond his own, for the "yeastiness of thought" that he covets. Jackson is always on the hunt for people who can offer him new ways to see the world or remind him of old ideas in a way that deepens his understanding. For Jackson, thinking is always a collective effort, and while he spends considerable time alone pondering questions, he is happiest in conversation.

My interaction with Jackson began in 2003, long after I had first begun reading his work, when he agreed to let me interview him about his ecological worldview. At the time, my activist energy was going into traditional left political work (antiwar organizing, the labor movement, feminist projects, and challenges to white supremacy), and I hoped that by publishing an interview on progressive websites I could spread Jackson's ecological thinking to people who were focused primarily on social justice. My first trip to Prairie Festival was in 2010, and I have been back every year since (except in 2020, when COVID-19 forced TLI to cancel the annual event). Once Jackson retired from running TLI and I retired from teaching at the University of Texas, our opportunities for engagement expanded, and since then I have done my best to keep up with Jackson's restless and relentless mind.

Keeping up means being open to Jackson's "spontaneous elaboration." He has long worried that educational institutions too often train students to "color within the lines," which can result in "graduates who are more given to minimal compliance than spontaneous elaboration." He tries to practice what he preaches, and the spontaneous elaborations just keep coming. In person, on the phone, or in correspondence, Jackson makes statements that—agree or disagree with him—demand attention.

Here's one Jackson elaboration that came spontaneously and stuck with me: "The world will always be more beautiful than useful." I take that to mean that while we rely on the larger living world around us for our sustenance, we should not ever see the world as merely a storehouse for humans, as the place we go just to stock up on supplies. To look at the world primarily in economic terms, focusing on what is useful to us, is to miss the deeper beauty—not just the things that make for pretty pictures but a beauty more powerful than our language can easily capture, a beauty that one feels without being able to explain.[1]

That comment from Jackson took me by surprise, in part because I tend to relegate beauty to the artists, assuming that it is a matter for poets and painters. Scientists and farmers are supposed to focus on practical matters, and so I assume they don't have much time for beauty. Jackson's comment reminded me of the false dichotomy I had unconsciously accepted.

"Even when people treat the world like a dump, all we need do is look up and look out, and it is better than any painting in any gallery," he continues.

Jackson walks through the world with eyes that constantly look up and out. On the wall of his office is one of his art-without-ego projects—a piece of ash that, when he cut it from a log, revealed a fascinating pattern inside the wood that appears to have been created by at least two different insects, one of which was likely the ash borer. He has collected enough of these pieces that the Birger Sandzén Memorial Gallery in Lindsborg, Kansas, mounted an "Art without

Ego" show in 2019. Jackson has nothing against paintings or other products of human creativity, but he has a deeper appreciation for the creativity of the ecosphere.

That might be as close as one can get to a Wes Jackson theology. Transcendence comes not from escaping the material plane of existence but from going deeper into the physical reality all around us. That is Jackson's path to loving the Creation without believing in a Creator. It is the way of a materialist mystic.

If these are contradictions, Jackson is not overly concerned. We use science in our attempt to learn about our location in the universe, and contradictions in science have to be resolved. But in the messy business of finding meaning in the universe, even if we steer clear of logical contradictions, conflict is inevitable—not just with others but within ourselves. For Jackson, one of those conflicts is between what he feels and what he knows. "I vacillate between my personality and my intellect," Jackson says. "I'm generally a happy, positive person, until I start thinking."

By that, Jackson means that an honest assessment of the state of the planet, and the extraordinary ways in which human societies have to change, is bound to lead a thinking person to consider the "end times"—not eschatological predictions about judgment day but a recognition that the political and economic systems that structure our world will not be around much longer. Unsustainable systems cannot be sustained. Unsustainable systems come to an end. Those end-of-systems times are coming.

Now in his eighties, Jackson doesn't necessarily expect to live to see the massive changes on the horizon, but he has grandchildren, and he can't help but worry about the world in which they will live out their lives. He thinks too much about soil erosion, chemical contamination, biodiversity loss, and climate disruption to "put on a happy face." Jackson's natural inclination is to smile, but smiling doesn't work like magic. "Why look so awfully tragic?" the song asks.[2] Because there is something tragic about the modern human condition.

Jackson often quotes Alfred North Whitehead on the subject: "The essence of dramatic tragedy is not unhappiness. It resides in the solemnity of the remorseless working of things."[3] Once we domesticated plants and animals, we became a species out of context, and from there humans giving in to the temptations of dense energy was simply the remorseless working of things. "The laws of physics are decrees of fate," Whitehead wrote, echoing a truism of the climate movement today—when politicians say we cannot change too quickly and must avoid disrupting the economy, environmentalists respond, "The laws of physics and chemistry don't negotiate." Humans' desire to squeeze more out of nature does not change the working of natural forces.

That is part of the tragedy in a world that remains beautiful. Jackson knows that there is no point in denying the consequences of ten thousand years of humans routinely giving in to that temptation. The "more joyful participation in the Creation" that he seeks will not come from ignoring reality. I suspect it is his appreciation for the beauty all around us that makes it possible for him to continue his work with such enthusiasm in the face of knowing what he knows about the ecological crises. An appreciation of the beauty, combined with the knowledge of how human recklessness can ignore that beauty, provides motivation to continue.

We have to face a growing list of "terrible truths," Jackson says, but he takes comfort from the English poet Kathleen Raine, who in an essay on T. S. Eliot wrote "that the statement of a terrible truth has a kind of healing power."[4] Once we face such truths there is no guarantee we can reverse course or repair all the damage done, but some kind of healing becomes possible. Is there any hope of healing our destruction of ecosystems if we have not yet confronted the truth of the destruction? Is rational planning to create more just and sustainable living arrangements possible without confronting the terrible truths?

Jackson certainly isn't overly optimistic about our prospects for a stable, decent future, but he cannot let go of hope. He resolves that conflict between what he knows and what he feels by holding on to what Albert Camus called a "stubborn hope":

> Tomorrow the world may burst into fragments. In that threat hanging over our heads there is a lesson of truth. As we face such a future, hierarchies, titles, honors are reduced to what they are in reality: a passing puff of smoke. And the only certainty left to us is that of naked

suffering, common to all, intermingling its roots with those of a stubborn hope.[5]

Wes Jackson is stubborn when defending what he loves, which includes, among many things, his view of the Smoky Hill River from the bluff behind his house; the work of TLI; the rolling prairie of the Flint Hills; his cattle; the old brick school building in Matfield Green; his family, friends, and neighbors; and people in general. There is beauty in it all, and he stubbornly holds on to the hope that all will endure and thrive.

Can the human species pull it off? Can we confront our carbon-seeking habits and embrace collectively imposed limits? Can we acknowledge our ignorance? Can we find our way back to the tree of life?

All of that is within human nature to do so long as we can control our human-carbon nature. We are smart enough to do it all as long as we do not overestimate our intelligence. We can commit to the task without knowing how it can be done.

What we do know is that nature does not care about us, or about any specific species for that matter. Whether or not we humans return to a right relation with the larger living world, natural forces will roll along—the world is not waiting for us to save it. Jackson likes to remind us that our relationship with nature is asymmetrical—we can love it, but it doesn't love us back. Our love will remain unrequited, but we are lost without that love.

Still, Jackson believes we owe it not only to ourselves but to the ecosphere to give it our best shot. While we humans are one small part, not the point, of the story of the universe, we also cannot ignore our distinctive place as storytellers.

"We are," Jackson says, "matter and energy's way of having gained self-recognition."

Appendix: Mission Statements of The Land Institute from The Land Report

"The Work of the Land Institute"

The Land Institute is devoted to a search for alternatives, alternatives in agriculture, energy, shelter and waste disposal. But it is more than that, for these alternatives bring to mind hardware, such stand-up equipment as wind generators and solar collectors. We are also devoted to a search for alternative world views, alternative thinking. Both have to come together, the alternative technology and the alternative thinking; for if we change the technology and still seek our identity by feeding our insatiable consumer-oriented appetites, we are still into mine depletion and sink impaction. At best, the earth's gain will be negligible, and our spirits, weary.

The way we have chosen to get ourselves in touch with the earth is to develop a holistic view and blend it with our work on specific problems and projects, oriented toward regional semi-self sufficiency. The holistic view can lead to such an amateurish approach to important subjects that one develops "opinions" or "views" on all subjects, indeed, on all matters of the day. Therefore, the appropriate counterweight is our involvement in specific projects. Hopefully, the holistic approach will help us avoid the usual pitfalls of the narrow specialist.

The Land Report, no. 1 (December 1976): 1, https://2hyzup3gkq37nm 98l33j3iwt-wpengine.netdna-ssl.com/wp-content/uploads/2018/06 /LR-1.pdf.

The Land Institute is a non-profit educational-research organization devoted to a search for sustainable alternatives in agriculture, energy, shelter and waste.

The Land Report, no. 3 (September 1977): 3, https://2hyzup3gkq37nm 98l33j3iwt-wpengine.netdna-ssl.com/wp-content/uploads/2018/06 /LR-3.pdf.

The Land Institute is a non-profit educational-research organization devoted to a search for sustainable agriculture and good stewardship of the Earth.

The Land Report, no. 34 (Fall 1988): 2, https://2hyzup3gkq37nm98l 33j3iwt-wpengine.netdna-ssl.com/wp-content/uploads/2018/06 /LR-34.pdf.

When people, land, and community are as one, all three members prosper; when they relate not as members but as competing interests, all three are exploited. By consulting Nature as the source and measure of that membership, The Land Institute seeks to develop an agriculture that will save soil from being lost or poisoned, while promoting a community life at once prosperous and enduring.

The Land Report, no. 53 (Summer 1995): 2, https://2hyzup3gkq37nm98l33j3iwt-wpengine.netdna-ssl.com/wp-content/uploads/2018/06/LR-53.pdf, currently online at https://landinstitute.org/about-us/vision-mission/.

Notes

Foreword

1. A chart created by the Defenders of Wildlife offers a more sophisticated version of that exercise: http://www.defenders.org/sites/default/files/publications/rural_crisis_matrix.pdf.
2. Stephen Heyman, *The Planter of Modern Life: Louis Bromfield and the Seeds of a Food Revolution* (New York: W. W. Norton, 2020), is a very illuminating portrait of a forgotten man worth remembering.
3. Marty Strange, *Family Farming: A New Economic Vision* (Lincoln: University of Nebraska Press, 1988).
4. Anne Case and Angus Deaton, *Deaths of Despair and the Future of Capitalism* (Princeton, NJ: Princeton University Press, 2020).

Introduction: In the Barn

1. Wes Jackson, Stan Cox, and Tim Crews, "The Next Synthesis," in Wes Jackson, *Nature as Measure: The Selected Essays of Wes Jackson* (Berkeley, CA: Counterpoint, 2011), 200.

2. Kernza® is the trademark for the intermediate wheatgrass (*Thinopyrum intermedium*) that is being developed at TLI. Perennial rice, called PR23, is a cross between an annual cultivated rice, *Oryza sativa*, and a perennial cousin of rice from Africa, *Oryza longistaminata*.

3. Wallace Stegner, *Beyond the Hundredth Meridian: John Wesley Powell and the Second Opening of the West* (New York: Houghton Mifflin, 1953), 256.

4. George Orwell, "Rudyard Kipling," *Horizon* (February 1942), http://orwell.ru/library/reviews/kipling/english/e_rkip.

5. Abraham J. Heschel, *The Prophets* (1962; repr., New York: Harper Collins, 2001), 19.

6. Appendix 2 includes the mission statements of TLI from its founding to today. See also "The Land Institute Collection," Kansas Historical Society, https://www.kshs.org/p/the-land-institute-collection/13779.

7. Wes Jackson, Aubrey Streit Krug, Bill Vitek, and Robert Jensen, "Transforming Human Life on Our Home Planet, Perennially," *Ecological Citizen* 2, no. 1 (2018): 43–46, https://www.ecologicalcitizen.net/pdfs/v02n1-08.pdf.

8. Craig Canine, "35 Who Made a Difference: Wes Jackson," Smithsonian.com, November 1, 2005, https://www.smithsonianmag.com/science-nature/35-who-made-a-difference-wes-jackson-114333178/.

9. Adam Gopnik, "Inquiring Minds: The Spanish Inquisition Revisited," *New Yorker*, January 16, 2012, https://www.newyorker.com/magazine/2012/01/16/inquiring-minds.

Chapter 1: The Energy of Life

1. Jackson borrows this boundary phrasing from Richard Levins and Richard Lewontin, *The Dialectical Biologist* (Cambridge, MA: Harvard University Press, 1985).
2. Jackson's interest in this idea appears in a poem by his friend Wendell Berry, "A Small Porch in the Woods," in *A Small Porch: Sabbath Poems 2014 and 2015* (Berkeley, CA: Counterpoint, 2016), 28.
3. Arnold Schultz, Ecosystemology Course Reader, 2009, https://nature.berkeley.edu/sites/default/files/Arnold%20Schultz%2C%20Ecosystemology%2C%202009.pdf.
4. Stan Rowe, "The Living Earth and Its Ethical Priority," 2003, http://www.ecospherics.net/pages/Roweliving.htm.
5. Steven A. Benner, "Defining Life," *Astrobiology* 10, no. 10 (December 2010): 1021–1030, doi: 10.1089/ast.2010.0524, https://www.ncbi.nlm.nih.gov/pmc/articles/PMC3005285/.
6. Luis P. Villarreal, "Are Viruses Alive?" *Scientific American*, August 8, 2008, https://www.scientificamerican.com/article/are-viruses-alive-2004/.
7. Heather Pringle, "The Ancient Roots of the 1%," *Science* 344, no. 6186 (May 23, 2014): 822–825, doi: 10.1126/science.344.6186.822, https://science.sciencemag.org/content/344/6186/822.
8. James C. Scott, *Against the Grain: A Deep History of the Earliest States* (New Haven, CT: Yale University Press, 2017).
9. Alfred W. Crosby, *Ecological Imperialism: The Biological Expansion of Europe, 900–1900*, 2nd ed. (Cambridge: Cambridge University Press, 2004); Jared Diamond, *Guns, Germs, and Steel: The Fates of Human Societies* (New York: W. W. Norton, 1997); and Ian Morris, *Why the West Rules—for Now: The Patterns of History, and What They*

Reveal about the Future (New York: Picador/Farrar, Straus and Giroux, 2010).

Chapter 2: Agriculture

1. David Montgomery, *Dirt: The Erosion of Civilizations*, 2nd ed. (Berkeley: University of California Press, 2012), xii.

2. National Oceanic and Atmospheric Administration, "NOAA Forecasts Very Large 'Dead Zone' for Gulf of Mexico," June 10, 2019, https://www.noaa.gov/media-release/noaa-forecasts-very-large-dead-zone-for-gulf-of-mexico.

3. "Ecosystems & Human Well-Being: Synthesis Report," Millennium Ecosystem Assessment, May 2005, https://www.millenniumassessment.org/documents/document.356.aspx.pdf.

4. US Environmental Protection Agency, "Sources of Greenhouse Gas Emissions," n.d., https://www.epa.gov/ghgemissions/sources-greenhouse-gas-emissions.

Chapter 3: Taking Nature Seriously

1. Mike Davis, *Late Victorian Holocausts* (London: Verso, 2001).

2. Alexander Pope, *Moral Essays*, "Epistle IV. Of the Use of Riches: To Richard Boyle, Earl of Burlington" (1731), https://www.bartleby.com/203/146.html.

3. Albert Howard, *An Agricultural Testament* (London: Oxford University Press, 1943).

4. "Historical Timeline—Farmers & the Land" (2014), https://www.agclassroom.org/ganarchive/timeline/farmers_land.htm; and

"Growing a Nation: An Interactive Timeline of the History of Agriculture in the United States," 2018, https://growinganation.org/.

Chapter 5: Knowledge

1. Wes Jackson, "Toward an Ignorance-Based Worldview," *The Land Report* no. 81 (Spring 2005): 14–16, https://2hyzup3gkq37nm98i33j3iwt-wpengine.netdna-ssl.com/wp-content/uploads/2018/05/101992-LR-81.pdf.
2. Bill Vitek and Wes Jackson, eds., *The Virtues of Ignorance: Complexity, Sustainability, and the Limits of Knowledge* (Lexington: University Press of Kentucky, 2008).
3. Wendell Berry, "The Way of Ignorance," in *The Virtues of Ignorance*, ed. Vitek and Jackson, 37.
4. Frank E. Egler, *The Nature of Vegetation, Its Management and Mismanagement: An Introduction to Vegetation Science* (Norfolk, CT: Aton Forest, 1977), quoted in Alan Haney and Rebecca L. Power, "Adaptive Management for Sound Ecosystem Management," *Environmental Management* 20, no. 2 (1996): 879.
5. For example, Stuart Firestein, *Ignorance: How It Drives Science* (New York: Oxford University Press, 2012).
6. Wingspread Statement on the Precautionary Principle (January 1998), http://www.sehn.org/precaution.html.
7. David Orr, "Technological Fundamentalism," *Conservation Biology* 8, no. 2 (June 1994): 335–337.
8. Mark Lynas, *The God Species: Saving the Planet in the Age of Humans* (Washington, DC: National Geographic Society, 2011).
9. Stewart Brand, "The Purpose," *Whole Earth Catalog*, https://

monoskop.org/images/0/09/Brand_Stewart_Whole_Earth_Catalog_Fall_1968.pdf.

10. Stewart Brand, *Whole Earth Discipline: An Ecopragmatist Manifesto* (New York: Viking Adult, 2009), 1.

11. Elisabeth Rosenthal and Andrew W. Lehren, "Relief in Every Window, but Global Worry Too," *New York Times*, June 20, 2012, http://www.nytimes.com/2012/06/21/world/asia/global-demand-for-air-conditioning-forces-tough-environmental-choices.html?pagewanted=all.

12. Stan Cox, a colleague of Jackson at TLI, explains how we can reduce our reliance on air-conditioning in *Losing Our Cool: Uncomfortable Truths about Our Air-Conditioned World (and Finding New Ways to Get through the Summer)* (New York: New Press, 2010).

Chapter 7: "We" Have to Accept Limits

1. Stan Cox, *The Green New Deal and Beyond: Ending the Climate Emergency While We Still Can* (San Francisco: City Lights Books, 2020).

Chapter 8: A Creaturely Worldview

1. Resilience.org, April 3, 2019, https://www.resilience.org/stories/2019-04-03/lets-get-creaturely-a-new-worldview-can-help-us-face-ecological-crises/.

2. Naomi Klein, *On Fire: The (Burning) Case for a Green New Deal* (New York: Simon and Schuster, 2019); and Robert Jensen, "The Danger of Inspiration: A Review of *On Fire*," CommonDreams.org, September 17, 2019, https://www.commondreams.org/views

/2019/09/17/danger-inspiration-review-fire-burning-case-green-new-deal.

3. William J. Ripple, et al., "World Scientists' Warning to Humanity: A Second Notice," *BioScience* 67, no. 12 (December 2017): 1026–1028, https://doi.org/10.1093/biosci/bix125.

4. Annie Corrigan, "Wendell Berry: Consciousness and the Creaturely Life," Indiana Public Media, February 27, 2011, https://indianapublicmedia.org/eartheats/wendell-berry-consciousness-creaturely-life.php.

5. Vaclav Smil, *Enriching the Earth: Fritz Haber, Carl Bosch, and the Transformation of World Food Production* (Cambridge, MA: MIT Press, 2000).

6. Stan Cox, "Renewables Alone Won't Save Us," *Dissent*, November 9, 2017, https://www.dissentmagazine.org/online_articles/100-percent-renewable-energy-overconsumption-inequality.

7. Stan Cox, *Any Way You Slice It: The Past, Present, and Future of Rationing* (New York: New Press, 2013).

8. Stan Cox and Paul Cox, "100 Percent Wishful Thinking: The Green-Energy Cornucopia," Counterpunch, September 14, 2017, https://www.counterpunch.org/2017/09/14/100-percent-wishful-thinking-the-green-energy-cornucopia/.

Conclusion: Spontaneous Elaboration

1. This borrows from Henry David Thoreau's 1837 observation that, "This curious world which we inhabit is more wonderful than it is convenient; more beautiful than it is useful; it is more to be admired and enjoyed than used." Thoreau, *The Writings of Henry David Thoreau*, ed. F. B. Sandborn (Boston: Houghton Mifflin, Co., 1893), 9.

2. Lee Adams and Charles Strouse, "Put on a Happy Face," from the musical *Bye Bye Birdie*, first performed in 1960.

3. A. N. Whitehead, *Science and the Modern World* (Cambridge: Cambridge University Press, 1932), 13.

4. Kathleen Raine, "The Poet of Our Time," in *T. S. Eliot: A Symposium*, ed. Richard March (New York: Tambimuttu & Mass, 1965), 79.

5. Albert Camus, "The Wager of Our Generation," in *Resistance, Rebellion, and Death* (New York: Vintage, 1960), 239–240.

Published Works of Wes Jackson

Books

Altars of Unhewn Stone: Science and the Earth. 1987; paperback repr., Wooster, OH: Wooster Book Co., 2006.
Becoming Native to This Place. 1994; paperback repr., Berkeley, CA: Counterpoint Press, 1996.
Consulting the Genius of the Place: An Ecological Approach to a New Agriculture. Berkeley, CA: Counterpoint Press, 2010.
Nature as Measure: The Selected Essays of Wes Jackson. Berkeley, CA: Counterpoint Press, 2011.
New Roots for Agriculture. Lincoln: University of Nebraska Press, 1980.

Edited Collections

Jackson, Wes, ed. *Man and the Environment*, 3rd ed. Dubuque, IA: W. C. Brown, 1979.

Jackson, Wes, Wendell Berry, and Bruce Colman, eds. *Meeting the Expectations of the Land: Essays in Sustainable Agriculture and Stewardship*. San Francisco: North Point Press, 1984.

Vitek, Bill, and Wes Jackson, eds. *The Virtues of Ignorance: Complexity, Sustainability, and the Limits of Knowledge*. Lexington: University Press of Kentucky, 2008.

Vitek, William, and Wes Jackson, eds. *Rooted in the Land: Essays on Community and Place*. New Haven, CT: Yale University Press, 1996.